DAVID VENNOR-MORRIS

MATHEMATICAL TOPICS

An Introduction to Vectors
SI Edition

MATHEMATICAL TOPICS

Other titles in this series

An Introduction to Vectors
SI Edition

A. E. Coulson A.R.C.Sc., B.Sc.

HEAD OF THE MATHEMATICS DEPARTMENT
GRAMMAR SCHOOL FOR BOYS, DOVER

Longmans

LONGMANS, GREEN AND CO LTD
London and Harlow
Associated companies, branches and representatives
throughout the world

© *A. E. Coulson 1967*
First published 1967

SI Edition 1969 © *Longmans, Green & Co Ltd 1969*

SBN 582 31753 3

Preface to the SI Edition

In this new edition the text has been enlarged by the addition of some new topics in order to bring the aims of the book in line with recent syllabus improvements, such as those introduced by many of the examining boards, and by the inclusion of those topics which develop the subject to greater advantage. This has also provided the opportunity of introducing many more exercises of varying difficulty, including some sets of original examples which carry the subject to a greater depth.

Although in its new expanded form this book will supplement many of the existing texts to meet the latest extensions of the new syllabuses in both pure and applied mathematics, in itself it still remains a complete introduction to the subject of vectors and continues to provide the student with a sound basic study of this increasingly important branch of mathematics.

SI units are used throughout although, for practical reasons, it has been felt necessary to use degrees as well as radians as units of angle.

Made and printed in Great Britain by
William Clowes & Sons Ltd, London and Beccles

Contents

vector – The angle between two intersecting straight lines – To find the perpendicular distance from a given point to a given line using the perpendicular vector – Parametric representation of the ellipse – Trajectory of a projectile – The circular helix

Chapter 1
Introduction

Historical development of the idea of vector quantities

From the time of Archimedes (250 B.C.) mathematicians have been able to compound forces and velocities to meet the requirements of navigation and engineering. In 1586 Simon Stevin of Bruges published treatments of statics and hydrostatics. He was the first mathematician of the sixteenth century to continue the work of Archimedes in statics and he used the triangle of forces to compound two forces; a method which he was the first to publish. A century later Sir Isaac Newton extended these ideas, introduced new concepts such as momentum, and stated three 'Laws', which enabled him to provide an axiomatic treatment based on these three assumptions. He was able to explain many known physical phenomena and to predict more. Within the mathematical framework built on his three assumptions he was able to explain astronomical facts then known and to formulate his theory of universal gravitation.

It should be emphasised that Newtonian Mechanics is still valid except when dealing with masses approaching those of the atom, and with velocities approaching that of light. Under these extreme conditions modifications, including quantum mechanics and relativity theory, become necessary. A century after Newton came great discoveries in physics which focused attention on the growing concept of Vector Quantities, the work of Gauss, Argand, Oersted, Ampère and Faraday. The discoveries of the last two men could only be explained in new mathematical ideas. The foundations had been laid, and in 1844 Sir W. R. Hamilton in Dublin and H. G. Grassmann in Stettin produced, independently of each other, an extension of algebraic ideas to explain known experimental results. Hamilton was the first mathematician to use the word VECTOR to represent this new concept, but he used the concept of vector within the wider concept of the quaternion and although his followers tried to use the theory of quaternions over an extensive field, the theory never became widely used and was eventually discarded. Grassmann's Theory of Extensions included the concept of the vector and the quaternion within an even broader framework; while Hamilton worked within real space of three

1

dimensions, Grassmann extended his ideas into an abstract space of *n*-dimensions.

At last the idea of a quantity which had magnitude and direction was recognised and given a name to distinguish it from a quantity with magnitude only, called a Scalar quantity. The following is a list of some of the more common Scalar and Vector quantities:

Vectors	*Scalars*
Displacements	Length
Velocity	Area
Force	Volume
Acceleration	Work done
Momentum	Electrical resistance
Electrostatic force	Power
Magnetic force	Energy
Electric current, etc.	Mass
	Density
	Specific gravity
	Temperature
	Potential, etc.

An examination of the table shows that the essential difference between the vector quantities and the scalar quantities is the additional quality they possess, besides magnitude, of *orientation* and *sense*, expressed by the one word DIRECTION. Throughout this book whenever the direction of a vector is mentioned these two aspects of direction will generally be intended.

During the second half of the nineteenth century efforts to use Hamilton's quaternions and Grassmann's *Ausdehnungslehre* in practical problems were not very successful, but many mathematicians in various countries were working to formulate a useful form of vector algebra and in 1881 Professor Willard Gibbs of Yale printed privately for the use of his students a pamphlet setting out the elements of a system of vector analysis which he had developed to handle practical problems in physics. Professor Gibbs had extracted from the works of Hamilton and Grassmann those methods and ideas which he could adapt to his treatment of vectors and it is mainly his treatment which has been developed into the modern method of vector algebra. In his original pamphlet Professor Gibbs pointed out that the simplest of all vectors is a straight line drawn from A to B, the length AB representing the magnitude, the direction of the line representing the direction of the vector with an arrowhead to give the sense. He pointed out that any vector could be represented by this geometrical model, which we call a directed line segment (although it was Stevin of Bruges who in 1586 first explained that a force could be represented by a line in magnitude and direction).

Free vectors, localised and bound vectors
Because vector quantities can be represented by directed line segments, a geometrical approach to vector algebra enables the student to build up abstract concepts from a practical basis. In order to make the best use of geometrical models we must start with FREE vectors. A free vector is one which has no specified point of application. We can imagine it capable of treatment in any convenient position. In the application of vectors to some practical problems there may be some restrictions on the position of the vector: for instance in mechanics a force vector is sometimes restricted to a certain line of action although it may be located anywhere along that line; such a vector is called a LOCALISED or SLIDING vector. In other problems it may be restricted to a certain position also, it is then said to be a BOUND vector. Throughout we shall be dealing only with free vectors, unless otherwise stated.

Geometrical representation of a vector
Suppose we are told that Ship A is travelling at 7 m/s in a NE direction and Ship B is travelling at 3 m/s in a NW direction. We can show these motions in a geometrical diagram.

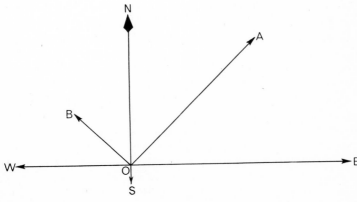

Fig. 1

The velocity of ship A is represented by the line segment **OA**
  (i) The length of OA is 7 units.
 (ii) The direction of OA is NE.
(iii) The arrowhead indicates that the motion is from O to A, i.e. the sense of the direction.
Point O is called the origin of the vector and point A is the terminus or end-point.
In a similar manner the vector **OB** represents the velocity of ship B. It is convenient to show both vectors with their origins at the same point since they are free vectors.

Chapter 2
Equal vectors

Coplanar vectors

Suppose now that two ships C and D are both travelling at 7 m/s in NE direction. Regardless of their actual positions obviously both have equal velocity vectors and if we represent them on a geometrical model we shall have the following:

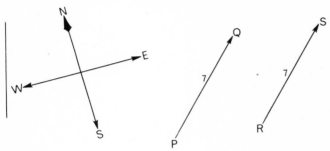

Fig. 2

PQ and **RS** will represent the two equal vectors and it is also obvious that if two vectors are equal then three conditions must hold:
(i) Vector **PQ** is parallel to vector **RS**.
(ii) **PQ** and **RS** have the same direction and the same sense.
(iii) The magnitude of **PQ** must be equal to the magnitude of **RS**.

In the two examples used so far, the vectors have been in the same plane because they have been assumed to be horizontal. The two vectors have been coplanar. We have been working in two dimensions only, which is simpler than working in three. In the first part of this book the treatment will be restricted to two dimensions, but at an appropriate stage the treatment will be extended into three-dimensional space.

Since a directed line segment can represent any vector quantity in magnitude, direction and sense, we can always make a geometrical model of a system of vectors. Displacements have magnitude, direction and sense, and constitute the simplest of all vectors; their representation by directed line segments is obvious. The physicist and the engineer live in a world dominated by vectors and the whole study of vectors started when Stevin in the course of his engineering work showed that a force can be represented by a line and then proceeded to show the geometrical method of adding forces by using the triangle of forces.

Algebraic representation
In addition to the geometrical representation of a vector we can use algebraic symbols on condition that they are suitably defined.

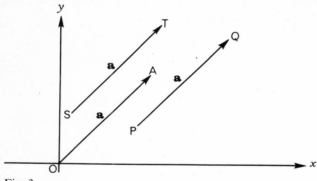

Fig. 3

In Figure 3 we show three coplanar vectors **PQ**, **ST** and **OA**, all with the same length or magnitude, all parallel and in the same sense. By our earlier statements these three conditions mean that the three vectors are all equal to each other. We can represent them all by the same algebraic symbol; in print this is shown by bold type, in handwriting or typewriting by underlining the letter. Any convenient letter can be used, in this case we use **a**. So we have **OA = PQ = ST = a**.

The magnitude of the vector is denoted by the use of two vertical lines thus |**a**| known as Modulus **a** or Mod **a**. In the diagram the two lines at right angles serve as the normal axes of reference and given a vector **a** we can conveniently show the starting point or origin of the vector at the intersection of the axes of reference, which is also called the origin in normal number algebra.

Suppose we are given two vector quantities **a** and **b** of the same kind, and then told that **a** = **b**, then it follows that **a** and **b** are equal in magnitude, parallel and in the same sense. The equality sign in vector algebra always has this threefold meaning.

Note also that the magnitude of a vector is always positive.

6

Addition of vectors

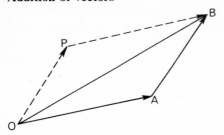

Fig. 4

Consider a displacement from O to A as shown. A body starting at O and ending at A suffers a displacement represented by the vector **OA**. If it then suffers another displacement from A to B represented by the vector **AB**, when it arrives at point B or position B, its total displacement, i.e. the SUM of **OA** and **AB**, is equivalent to the displacement from O to B. This is shown by **OA**+**AB** = **OB**. The equality sign has the threefold meaning already given to it earlier and the plus sign in vector algebra has this new meaning of geometrical addition using the Triangle of Vectors.

If **OA** = **a**, **AB** = **b** and **OB** = **c**, then we can represent the process of addition of **a** and **b** as

a+**b** = **c**.

In the diagram if **OP** is equal and parallel to **AB** then they are equal vectors, **PB** is equal and parallel to **OA** hence **OP** = **b** and **PB** = **a** and from triangle OPB we have

b+**a** = **c**.

It follows from this that **a**+**b** = **b**+**a** and in vector algebra we have shown that addition is commutative.

The addition of vectors of the same kind is defined by the Triangle Rule and agrees with the experience of engineers and physicists. If two forces A and B act at a point in a body then the total effect is exactly equal to that which would result from a single force called the resultant. The two forces A and B could be replaced by a single force C, which by experiment is found to be equal to the force C predicted by vector methods as the vector sum of A and B. Over the centuries this has been one of the fundamental theorems of mechanics and in fact the mathematical treatment of vectors is justified by its agreement with physical results.

Worked examples on vector addition
1. On a vector diagram represent the following vectors:
    (i) A force of 2 N making an angle of 30° with the horizontal direction.
    (ii) A force of 3 N making an angle of 70° with the horizontal.
    (iii) The sum of these two vectors.
    Scale: let 2 cm represent 1 N.

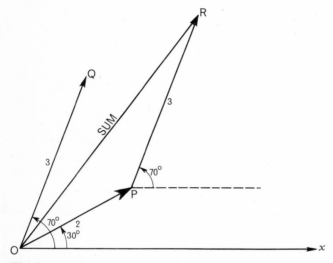

Fig. 5

OP represents the force of 2 N in a direction making 30° with O*x* the horizontal.

OQ represents the force of 3 N at an angle of 70° to O*x*. From P draw PR equal and parallel to OQ, then **PR** also represents the force of 3 N.

OR then represents the SUM of **OP** and **PR** and is 4·75 N at an angle of 55° or 0·96 rad with O*x*.

2. A yacht capable of sailing at 7 km/h is heading in a NE direction but a tide of $2\frac{1}{2}$ km/h is running towards N 20° W. Find the actual speed and direction of the yacht.

Let O be the origin, from O draw OP 7 cm long in a direction N 45° E. **OP** then represents the velocity due to the yacht. From P draw PQ $2\frac{1}{2}$ cm long in a direction N 20° W, then **OQ** is the resultant velocity of the yacht. The speed of the yacht is approximately 8·4 km/h in a direction N $29\frac{1}{2}$° E.

8

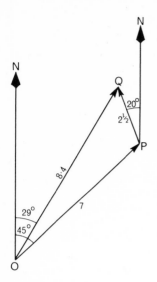

Fig. 6

Exercise 1.

1. An airplane capable of 250 km/h in still air flies NW at this speed and meets a wind of 45 km/h from the SW. Find from a vector diagram the final speed and direction of the aircraft.

2. A ship steaming at 16 km/h on a bearing 125° runs into a stream flowing at 2 km/h in a direction due East. Find the speed and direction of the actual movement of the ship.

3. Two forces of 6 N making an angle of 60° with each other act on a body. What is the resultant force on the body?

4. A rocket falling with an acceleration of 9·81 m/s² experiences a side thrust causing a horizontal acceleration of 0·92 m/s². What is the resultant acceleration and the angle it makes with the vertical direction?

5. An aircraft capable of 250 km/h in still air sets a course NE but its track is actually N 40° E and its ground speed is 260 km/h. Find the speed and direction of the wind that caused this deviation from the set course.

6. A ship A starts from a point O and travels 3 km North then 4 km NE. Another ship B also starts from O but travels 4 km NE and then 3 km North. Where are both ships finally? Show this on a vector diagram and give your conclusion.

Subtraction of vectors

In vector algebra a special meaning attaches to the symbols + and =. We now proceed to find a meaning for the negative sign and a meaning for vector subtraction.

Early in a school course students learn that the minus sign can have two meanings depending on the branch of mathematics in which they are working. In arithmetic the sign means subtract, e.g. $7-3$ means subtract 3 from 7, result 4. But $3-7$, i.e. subtract 7 from 3, is impossible until the student has learnt the meaning of DIRECTED numbers, then -7 no longer means subtract 7 but 7 units in a direction opposite to $+7$. -7 is called a directed number and can be shown on the number line.

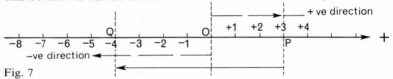

Fig. 7

Using a horizontal line to represent the number scale and taking direction to the right as positive and direction to the left as negative, we start at 0 and take 3 positive steps to the position P and then take 7 negative steps *from* P in the opposite direction, which then brings us to Q at the point -4. Thus using directed numbers we now state that $+3-7 = -4$. This could be restated as $+3+(-7) = -4$, or in words to three positive steps add 7 steps in the opposite negative direction, which results in 4 negative steps. In the same way $+3+(-3) = 0$ and so generally $+n+(-n) = 0$ in number algebra.

Similar ideas produce similar results in vector algebra. In diagram (8) if **OP** represents vector **p**, then **OP′** of the same magnitude, of the same direction but opposite in *sense* is defined as $-\mathbf{p}$.

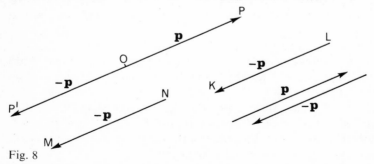

Fig. 8

Since vector **NM** and vector **LK** are parallel to **OP′**, equal in magnitude and have the *same sense*, then they are both equal to **OP′**. They are also equal to −**p**. Notice that |−**p**| = |**p**|.

Now let vector **r** of very small magnitude be equal to the vector sum of **a** and **b** as shown in the diagram.

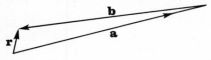

Fig. 9

By the triangle rule for the addition of vectors, **r** = **a**+**b**, and as |**r**| becomes very small, vector **b** will approach vector **a** in magnitude but will tend to be in the opposite direction. As |**r**| → 0, |**b**| → |**a**| and will be in the opposite sense, thus **b** → −**a**.

If the process is continued until |**r**| = 0, we say that vector **r** has become the ZERO VECTOR and then

a+**b** = **0**

But

b = −**a**

∴   **a**+(−**a**) = **0**

As in number algebra, this is usually written

a−**a** = **0**.

As with number algebra, we can now define the subtraction of vector **b** from vector **a** as the addition of −**b** to **a**.

a−**b** = **a**+(−**b**)

In the vector diagram we show this process

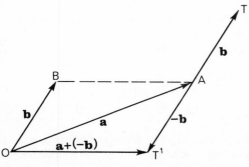

Fig. 10

2

If **OB** and **OA** represent the two vectors **b** and **a**, through the terminal point A draw TAT′ parallel to **OB**. Let |AT| = |OB| in length: then since **AT** is equal, parallel to and in the same sense as **OB**, **AT** represents vector **b**. If |AT′| = |AT| then **AT′** is the vector −**b**. **OT′** is the vector sum of **a** and −**b**.

OT′ = **a** + (−**b**)

= **a** − **b**.

In Figure 10 above it is seen that |**BA**| is equal to |**OT′**|, parallel and in the same sense. It also represents the vector **a** − **b**. This suggests that in general to find the difference of two vectors we need only set them out as the adjacent sides of a triangle but with a common origin and the third side will be the difference of the two vectors. To find the difference of vectors **p** and **q** we proceed as follows:

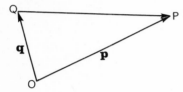

Fig. 11

If **OP** and **OQ** are the two vectors **p** and **q**, then **QP** is **p** − **q** because a displacement from Q to P is equivalent to a displacement from Q to O followed by another from O to P, or **QP** = **QO** + **OP** but **QO** = −**q** hence

QP = −**q** + **p**

= **p** − **q**.

The *sense* of **QP** must be emphasised at this point because **PQ** = −**QP** and using the method of the previous paragraph we have **PQ** = **PO** + **OQ** and **PO** = −**OP** = −**p**

hence **PQ** = −**OP** + **OQ**

= −**p** + **q**

= **q** − **p**.

We have also shown that in vector algebra we can operate on vector equations with the negative sign as in number algebra because we have shown that **PQ** = −**QP** = −(**p** − **q**) = **q** − **p**.

The parallelogram of vectors

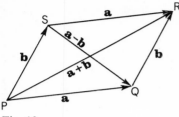

Fig. 12

If the sides of the parallelogram PQRS, **PQ** and **PS** represent the vectors **a** and **b**, then **SR** and **QR** also represent the vectors **a** and **b**. The diagonal **PR** = **a** + **b** and the diagonal **SQ** = **a** − **b** since
PR = **PQ** + **QR** and
SQ = **SP** + **PQ** = − **PS** + **PQ**.

In mechanics the sum or resultant of two forces is often found by using the Parallelogram of Forces.

The polygon of vectors

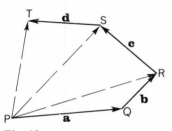

Fig. 13

When more than two vectors are to be dealt with they can be added or subtracted two at a time. If we wish to find **a** + **b** + **c** + **d** we draw **PQ** to represent **a**, then from the terminal point of **PQ** we draw **QR** to represent **b**, from the terminal point of **QR** we draw **RS** to represent **c** and similarly **ST** to represent **d**.

PR = **a** + **b** and in the triangle PRS,
PS = **PR** + **RS** ⇒ **PS** = **a** + **b** + **c**

In the triangle PST, **PT** = **PS** + **ST** ⇒ **PT** = **a** + **b** + **c** + **d**

Any number of vectors can be summed in this manner, using the polygon of vectors. Using the same vectors, if we wish to find **a** + **b** − **c** − **d** the diagram would be set out with the **RS**^{1} = − **RS** and **S**^{1}**T**^{1} = − **ST**

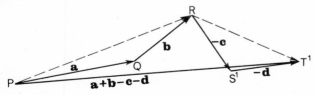

Fig. 14

PT^1 = **a** + **b** − **c** − **d** (i)

13

It is shown by the figure that $\mathbf{PT^1} = \mathbf{PR} + \mathbf{RT^1}$ and $\mathbf{PR} = \mathbf{a} + \mathbf{b}$ but
$\mathbf{RT^1} = -\mathbf{T^1R} = -(\mathbf{T^1S^1} + \mathbf{S^1R})$

$$= -(\mathbf{d} + \mathbf{c})$$
$$= -(\mathbf{c} + \mathbf{d}).$$

The final result can therefore be expressed also as

$$\mathbf{PT^1} = \mathbf{a} + \mathbf{b} - (\mathbf{c} + \mathbf{d}) \qquad \text{(ii)}$$

Combining equations (i) and (ii) gives
$$\mathbf{a} + \mathbf{b} - \mathbf{c} - \mathbf{d} = \mathbf{a} + \mathbf{b} - (\mathbf{c} + \mathbf{d}).$$

Once more vector equations have been shown to follow the same rules for signs and brackets as the equations of number algebra.

Relative velocity
The method of finding the difference of two vectors finds a useful application in problems on Relative Velocity. A motorist A travelling at 60 km/h is overtaking another motorist B travelling at 45 km/h along a straight road. To the motorist B the other car appears to be overtaking at 15 km/h. This is a matter of common experience. To the motorist A the other car appears to be travelling in the opposite direction at 15 km/h. To find the relative velocity of the other car the observer subtracts *his velocity* from the velocity of the other. So to B, A appears to be travelling at $+15$ km/h, but to A, B appears to be travelling at -15 km/h, i.e. in the opposite direction.

Meaning of the negative sign in vector algebra
At this stage it will be useful to remind the student that we have now given *three* different meanings to the negative sign:
1. The simple meaning as in arithmetic of subtracting one cardinal number from another;
2. The further idea of a directed number meaning a distance along the number line in the opposite direction to $+a$;
3. The vector meaning of compounding by the triangle or parallelogram method a vector equal in magnitude but opposite in direction to vector $\mathbf{a}$.

Which meaning you attach to the minus sign depends on the algebra in which you are working.

For those familiar with matrix algebra still another meaning is 'assigned' to the minus sign.

Example

Two aircraft A and B are flying at the same altitude but on different courses. Aircraft A is flying at 250 km/h towards the NE, and aircraft B is flying at 150 km/h towards the NW. What is the velocity of B relative to A?

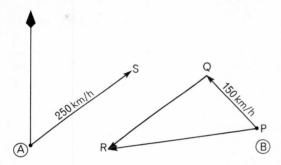

Fig. 15

To an observer on the ground the two aircraft can be observed on their separate courses at the same time, but to the pilot of A the movement of B is relative to his own aircraft and we have shown that in this case the velocity of A must be subtracted from the velocity of B to give the required relative velocity.

If **PQ** is the displacement vector representing 150 km/h to the NW, then **QR** represents the velocity opposite in direction but equal in magnitude to the velocity of A. Hence **PR** represents the velocity of B relative to A which is 292 km/h in a direction S 76° W. To the pilot of aircraft B the velocity of aircraft A relative to him is equal in magnitude to this but opposite in direction, i.e. 292 km/h in a direction N 76° E.

Exercise 2.

1. A yacht is travelling at 10 km/h in a northerly direction and a motorboat is sailing NW at a speed of $12\frac{1}{2}$ km/h. What is the velocity of the motorboat relative to an observer on the yacht? What is the velocity of the yacht relative to the helmsman of the motorboat?

2. A motorist M is approaching a crossroads from the West at 50 km/h on a level road and another motorist N is approaching the same crossroads from the South at 40 km/h on the same level. What is the velocity of M relative to N?

15

3. Heavy snow is blowing at 15 km/h from the NE and a cyclist is riding at 12 km/h in NW direction. What appears to be the velocity of the snow to the cyclist?

4. A motorist is travelling at 55 km/h in a northerly direction and rain is blown by the wind from the SE at 20 km/h. How does the rain appear to be blowing to the motorist?

5. A motorist makes a journey on a straight road running due North and South. On the outward journey to the North the wind appears to be blowing from the NE, when his speed is 30 km/h. On the return journey, when his speed is 35 km/h, the wind appears to be blowing from the SE. What is the true speed and direction of the wind if it remains constant throughout the journey?

Multiplication of a vector quantity by a scalar number

In number algebra, $3x$ means 3 times x; the number represented by x is multiplied by the number 3. From the work already done on vectorial addition we shall find a meaning for $3\mathbf{a}$ in vector algebra.

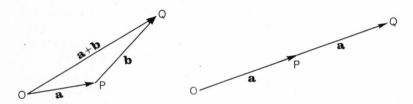

Fig. 16

If the displacement vector **OP** represents the vector quantity **a** and the displacement vector **PQ** represents vector quantity **b** then **OQ** = **OP**+**PQ** = **a**+**b**.

If now the vector **PQ** is altered so that |**b**| = |**a**| and the direction is the same as that of **OP**, then **b** becomes equal (in the vector meaning) to **a**. Since **OQ** = **OP**+**PQ** and both **OP** and **PQ** are equal to **a**, then **OQ** = **a**+**a**.

In number algebra, $x+x = 2x$; similarly in vector algebra under the conditions in the diagram, **OQ** = 2**a** = **a**+**a**. The diagram shows the meaning that we give to 2**a**; it represents a vector in the same direction (including sense) as vector **a** and having twice the magnitude. Similarly, 3**a** means a vector in the same direction as vector **a** but having three times the magnitude. By extending the process we can define $n\mathbf{a}$, when n is a positive real number, as a vector having the same direction and sense as vector **a** but with the magnitude n times that of **a** i.e.

$|na| = n|a|$.

Notice also that na is still a vector. The product of a scalar and a vector is a vector quantity.

If n is a NEGATIVE real number then the product na is now a vector in the opposite sense to vector **a** although in the same direction and since the magnitude of any vector is always positive then the magnitude will be the positive value of $|na|$, as would be understood in number algebra by the Modulus sign. In the diagram we again show the vector **a** and vector -2**a**.

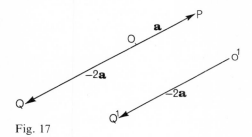

Fig. 17

OP represents vector **a** and **OQ** therefore represents vector -2**a**. Notice that since O^1Q^1 is equal and parallel to **OQ** it also represents -2**a**.

If $n = 0$, then the vector na is called the ZERO vector, its magnitude is zero and its direction and sense are not defined. The zero vector is denoted by **O**.

The meaning of m(na) when m and n are both scalars
We have shown that na is a vector whose magnitude is n times that of **a** but having the same direction and sense as the vector **a**. If we put **b** $= na$ then $m(na) = mb$. Hence $m(na)$ has the same sense and direction as **b** but m times the magnitude of **b**, and in its turn **b** has the same sense and direction as vector **a** but n times the magnitude, giving the result that $m(na)$ is a vector with the same direction and sense as **a** but with mn times the magnitude.

$|m(na)| = mn|a|$

$\qquad = nm|a|$ since m and n are scalar numbers.

17

To show the meaning of $(n+m)a$

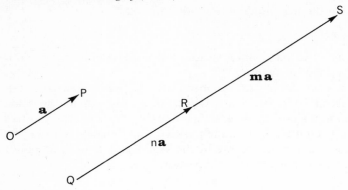

Fig. 18

If **OP** represents vector **a**, then n**a** and m**a** will have the same direction as **OP**. If **QR** represents n**a** and **RS** represents m**a** then **QR** and **RS** have the same direction and a common point **R**. **QRP** must therefore be in the same straight line.

Since **QRS** is a straight line

$$QS = QR + RS$$

$$= n\mathbf{a} + m\mathbf{a}.$$

But line QS is $(n+m)$ times the length of line OP and it follows that $\mathbf{QS} = (n+m)\mathbf{OP}$

$$= (n+m)\mathbf{a}$$

then

$$(n+m)\mathbf{a} = n\mathbf{a} + m\mathbf{a}.$$

This is the Distributive Law for the Scalar Multiplication of a Vector.

Multiplication of two vectors by a scalar

OP represents vector **a**, **PQ** represents vector **b** and **OQ** is therefore **a** + **b**. If OP is produced to R so that OR is n times OP, then **OR** is the vector n**a**. From R, RS is drawn parallel to PQ and n times its length and is therefore n**b**. This line OS is n times the length of

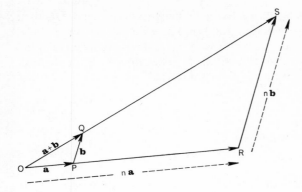

Fig. 19

the line OQ and parallel to it by the properties of parallel lines. This means that $\mathbf{OS} = n\mathbf{OQ}$ but

$$\mathbf{OS} = n\mathbf{a} + n\mathbf{b}$$

so we have

$$n\mathbf{OQ} = n\mathbf{a} + n\mathbf{b}$$

$$n(\mathbf{a} + \mathbf{b}) = n\mathbf{a} + n\mathbf{b}$$

This is also one of the Distributive Laws of Scalar Multiplication of vectors and corresponds to a similar law in number algebra.

The unit vector

Throughout the whole treatment of vectors (past as well as present) the idea of the UNIT VECTOR has been of the greatest assistance. As its name implies the unit vector is a vector of unit magnitude in any direction and sense which is under consideration.

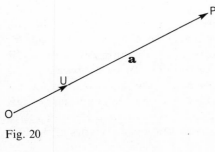

Fig. 20

2*

Suppose **OP** represents vector **a** of magnitude |**a**|, and **OU** is a vector in the same direction and sense as **a** but of unit magnitude. In this case since the unit vector is in the direction of **a** it is denoted by **â**. Later in different circumstances other notations for the unit vector will be used, especially when it is used independently of other vectors.

$$|\mathbf{OP}| = |\mathbf{a}|$$

$$|\mathbf{OU}| = |\mathbf{â}| = \text{unit magnitude}$$

$$\Rightarrow \mathbf{OP} = |\mathbf{a}|\mathbf{OU} = |\mathbf{a}|\mathbf{â}.$$

We distinguish between **a** and the unit vector in direction of **a** by calling **â**—CAP **a**.

Any vector can be represented in this way as the product of its magnitude and a unit vector in the same sense and direction as itself. (It must be remembered that the magnitude of a vector is a scalar so that the product of the magnitude and the unit vector remains a vector.)

A unit vector in any direction is often denoted by **u**. Let this unit vector **u** make an angle θ with a certain direction which will be called the BASE direction.

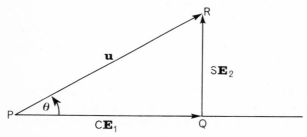

Fig. 21

If $\mathbf{E_1}$ is the unit vector in the base direction and $\mathbf{E_2}$ is the unit vector in the direction perpendicular to the base direction, $\mathbf{E_1}$ and $\mathbf{E_2}$ are called the base vectors. In the diagram $\mathbf{PR} = \mathbf{PQ} + \mathbf{QR}$ and we have just shown that vector **PQ** can be expressed as the magnitude of **PQ** times the unit vector in the same direction, i.e. $\mathbf{E_1}$. If the magnitude of **PQ** is denoted by c and the magnitude of **QR** is denoted by s then $\mathbf{PQ} = c\mathbf{E_1}$ and $\mathbf{QR} = s\mathbf{E_2}$ and the unit vector $\mathbf{u} = c\mathbf{E_1} + s\mathbf{E_2}$. $c\mathbf{E_1}$ and $s\mathbf{E_2}$ are the Component vectors of the unit vector, and c and s are scalars. Whatever the size of the angle the scalars c and s are always less than unity, and the scalar numbers c and s can be defined as the cosine and sine respectively of the angle θ between the unit vector and the base direction chosen.

Since the unit vector $\mathbf{E_2}$ is at right angles to the Base direction, $\mathbf{E_1}$ and $\mathbf{E_2}$ denote an orthogonal Base. Now while $c\mathbf{E_1}$ and $s\mathbf{E_2}$ are called the orthogonal component *vectors* of the unit vector, the *scalars* c and s are often referred to simply as the *components*. We will now consider a vector $\mathbf{a}$ in the direction and sense of this unit vector $\mathbf{u}$. Let $\mathbf{PA} = \mathbf{a}$.

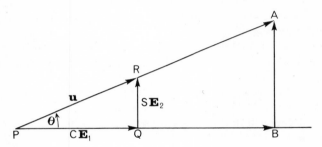

Fig. 22

Since $\mathbf{a}$ and $\mathbf{u}$ have the same sense and direction, $\mathbf{a} = |\mathbf{a}|\mathbf{u}$. But $\mathbf{u} = c\mathbf{E_1} + s\mathbf{E_2}$ which means that

$$|\mathbf{a}|\mathbf{u} = |\mathbf{a}|(c\mathbf{E_1} + s\mathbf{E_2})$$

$$= |\mathbf{a}|c\mathbf{E_1} + |\mathbf{a}|s\mathbf{E_2}$$

Hence we can express the vector $\mathbf{a}$ in terms of the base vectors,

$$\mathbf{a} = |\mathbf{a}|c\mathbf{E_1} + |\mathbf{a}|s\mathbf{E_2}.$$

As before the scalars $|\mathbf{a}|c$ and $|\mathbf{a}|s$ are called the *components* of the vector $\mathbf{a}$ in the orthogonal base. When the unit base vectors $\mathbf{E_1}$ and $\mathbf{E_2}$ are multiplied by the scalars $|\mathbf{a}|c$ and $|\mathbf{a}|s$ respectively the resulting vectors $\mathbf{PB}$ and $\mathbf{BA}$ are the component vectors of the vector $\mathbf{a}$.

$$\mathbf{PB} = |\mathbf{a}|c\mathbf{E_1} \text{ or } |\mathbf{a}| \cos \theta \, \mathbf{E_1}$$

and

$$\mathbf{BA} = |\mathbf{a}|s\mathbf{E_2} \text{ or } |\mathbf{a}| \sin \theta \, \mathbf{E_2}$$

21

Exercise 3.

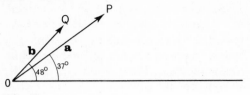

Fig. 23

Vector **a** has magnitude 3 units in the direction shown and vector **b** has magnitude 2 units in the direction OQ.

1. With the vectors given, construct vector **c** = **a**+**b**. Give the magnitude and direction of **c**.

2. Again with the given vectors construct **d** = 3**a**+3**b** and give the magnitude and direction of **d**. What is the relation between vectors **c** and **d**?

3. Mark OU = 1 unit along OP. **OU** therefore represents the unit vector along vector **a** and **OU** = **â**. Express **a** in terms of **â**.

4. Show the unit vector **b̂** in the diagram and express **b** in terms of the unit vector **b̂**.

5. Give **c** and **d** each in terms of the appropriate unit vector.

6. Choose a suitable unit and show a force of 3 N acting in a direction N 30° E and a force of 4 N acting in a direction S 30° E. Find the resultant (i.e. the sum) of these two forces.

The addition of three vectors

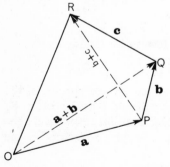

Fig. 24

Consider the addition of vectors **a**, **b** and **c** represented in the diagram above by **OP**, **PQ** and **QR** respectively.

$$\mathbf{OQ} = \mathbf{OP}+\mathbf{PQ} = \mathbf{a}+\mathbf{b}$$

$$\mathbf{PR} = \mathbf{PQ}+\mathbf{QR} = \mathbf{b}+\mathbf{c}$$

The vector **OR** can be obtained in two ways, from triangle OPR or from triangle OQR.

$$\mathbf{OR} = \mathbf{OP}+\mathbf{PR} = \mathbf{a}+\mathbf{PR} = \mathbf{a}+(\mathbf{b}+\mathbf{c})$$

and

$$\mathbf{OR} = \mathbf{OQ}+\mathbf{QR} = (\mathbf{a}+\mathbf{b})+\mathbf{QR} = (\mathbf{a}+\mathbf{b})+\mathbf{c}.$$

We have thus shown that

$$\mathbf{a} + (\mathbf{b} + \mathbf{c}) = (\mathbf{a} + \mathbf{b}) + \mathbf{c}.$$

This is the Associative Law for the Addition of Vectors. The curved brackets ( ) have the same meaning in vector algebra as in number algebra and matrix algebra, i.e. the operation between the brackets is to be performed first.

The addition of more than three vectors is performed in a similar manner by extending the process, *any* pair of vectors can be added and replaced by a single vector (their sum) and by successive reduction in pairs, the vectors can all be added. This was demonstrated earlier but the Associative Law for Addition shows that the *order* in which the vector addition is carried out is not important.

Vector equations

If $\mathbf{a} + \mathbf{b} = \mathbf{c}$ then the equality still holds if the vector $(-\mathbf{b})$ is added to both sides of the equation.

$$(\mathbf{a} + \mathbf{b}) + (-\mathbf{b}) = \mathbf{c} + (-\mathbf{b}) = \mathbf{c} - \mathbf{b} \tag{i}$$

But applying the Associative Law it follows that

$$(\mathbf{a} + \mathbf{b}) + (-\mathbf{b}) = \mathbf{a} + (\mathbf{b} + (-\mathbf{b}))$$

$$= \mathbf{a} + (\mathbf{b} - \mathbf{b})$$

$$= \mathbf{a}.$$

Since it was shown earlier that $\mathbf{b} - \mathbf{b} = \mathbf{0}$ the zero or null vector. So that equation (i) can now be written

$$\mathbf{a} = \mathbf{c} - \mathbf{b}.$$

We have now established that if $\mathbf{a} + \mathbf{b} = \mathbf{c}$ then $\mathbf{a} = \mathbf{c} - \mathbf{b}$. Once more we have shown that in vector algebra equations can be handled as in number algebra as far as addition, subtraction and scalar multiplication are concerned.

Suppose $\mathbf{x} + \mathbf{a} = \mathbf{b}$ where $\mathbf{a}$ and $\mathbf{b}$ are known *vectors* and $\mathbf{x}$ is an unknown *vector*, this constitutes a vector equation similar to the usual type of number equation $x + a = b$, where a and b are known numbers and x is the unknown number. We have shown that this can be rewritten $\mathbf{x} = \mathbf{b} - \mathbf{a}$, hence the equation has been solved in a vectorial sense.

If **a** and **b** are non-parallel, let $(x+y)\mathbf{a}+(x-y)\mathbf{b} = \mathbf{a}$, where x and y are scalars. The lefthand side of this equation is the sum of two vectors and can be expressed as a single vector but then the equality is only possible if the vector **b** is the zero vector. This means that

$$x - y = 0$$
$$\Rightarrow x = y$$

The equation then reduces to

$$(x+y)\mathbf{a} = \mathbf{a}$$
$$\Rightarrow (x+y) = 1$$
$$\Rightarrow x = \tfrac{1}{2} \text{ and } y = \tfrac{1}{2}.$$

If **a** and **b** are non-parallel vectors and $x\mathbf{a}+y\mathbf{b} = 0$ then both x and y must be zero.

Suppose that $x \neq 0$ then the equation can be rearranged

$$x\mathbf{a} = -y\mathbf{b}$$
$$\mathbf{a} = -(y/x)\mathbf{b}.$$

But this means that vectors **a** and **b** are parallel to each other and this is contrary to the given statement, hence $x = 0$. If $x = 0$ then it follows that $y = 0$. Under these conditions the two vectors **a** and **b** are said to be Linearly Independent.

Parallel vectors

By definition, if two vectors **a** and **b** are *equal* they are equal in magnitude, have the same direction and in the same *sense*, i.e. they are parallel. Since we are dealing with free vectors we must be prepared to think of a vector being parallel to itself (see page 6).

If $\mathbf{c} = \tfrac{1}{2}\mathbf{d}$ then the two vectors **c** and $\tfrac{1}{2}\mathbf{d}$ being equal means that **c** is parallel to $\tfrac{1}{2}\mathbf{d}$, and from the definitions on page 16, $\tfrac{1}{2}\mathbf{d}$ is parallel to **d**, so we can conclude that vector **c** is parallel to vector **d**, has the same sense but only half the magnitude. This conclusion is not dependent on their starting points and is true for any number of dimensions.

Fig. 25

Since $-\mathbf{a}$ is defined as a vector equal in magnitude to vector $\mathbf{a}$ parallel to it *but* opposite in sense, then $\mathbf{c} = -\frac{1}{2}\mathbf{d}$ means that $\mathbf{c}$ is parallel to $\mathbf{d}$ but has half the magnitude of $\mathbf{d}$ and is in the opposite sense.

Development of vector algebra
It is convenient at this stage to summarise the work so far which has been used to develop an algebra of vectors. The familiar signs $=$, $+$, $-$ have special meaning when used in vector algebra:
The equality sign $(=)$ means having the same magnitude and the same direction (including sense).
The addition sign $(+)$ means vectorial addition, i.e. by the triangle of vectors.
The negative sign $(-)$ means a vector having the same magnitude and direction but the opposite sense to a positive vector, so that we write for $\mathbf{a} - \mathbf{a} = 0$, $\mathbf{a} + (-\mathbf{a}) = \mathbf{0}$.

The only kind of multiplication dealt with so far has been the multiplication of a vector by a scalar number and we must be very careful to show this in a manner which does *not* use any special sign or symbol since special meanings will be developed later on for the more familiar multiplication signs used in number algebra. The product of a scalar number m and the vector $\mathbf{a}$ is simply written as $m\mathbf{a}$ which is understood to mean a vector with magnitude m times that of vector $\mathbf{a}$ but having the same direction. If m is a positive number then the sense is the same as that of $\mathbf{a}$, but if m is a negative number the sense is opposite. In addition we shall only consider m when it is a *real* number, and further, division by the scalar real number m will be understood as in number algebra to be equivalent to multiplication by the reciprocal of m, i.e. $1/m$.

Summary of laws
With the meanings for the symbols set out above we can now list the Laws of Vector Algebra established so far in the treatment of vectors in a plane (sometime known as 2-vectors).
1. $\mathbf{a} + \mathbf{b} = \mathbf{b} + \mathbf{a}$ Commutative Law for Addition.
2. $\mathbf{a} + (\mathbf{b} + \mathbf{c}) = (\mathbf{a} + \mathbf{b}) + \mathbf{c}$ Associative Law for Addition.
The curved brackets have the same meaning as in number algebra, i.e. it is intended that the addition inside the brackets shall be performed first, the brackets indicate *order* of performing the operation.

3. $m\mathbf{a} = \mathbf{a}m$ Commutative Law for Scalar Multiplication.
4. $m(n\mathbf{a}) = (mn)\mathbf{a}$ Associative Law for Scalar Multiplication.
5. $m(\mathbf{a}+\mathbf{b}) = m\mathbf{a}+m\mathbf{b}$⎰ Distributive Laws for Scalar
6. $(m+n)\mathbf{a} = m\mathbf{a}+n\mathbf{a}$ ⎰   Multiplication.

A UNIT VECTOR is a vector having UNIT MAGNITUDE. A unit vector in the direction and sense of vector $\mathbf{a}$ is usually denoted by $\hat{\mathbf{a}}$ to indicate this special property and $\mathbf{a} = |\mathbf{a}|\hat{\mathbf{a}}$.

If $\mathbf{a} = \mathbf{b}$ then $\mathbf{a}-\mathbf{b}$ is defined as the ZERO VECTOR or NULL VECTOR; it has zero magnitude and its direction is *indeterminate*.

If $\mathbf{a} = m\mathbf{b}$ since $m\mathbf{b}$ is also a vector equal to $\mathbf{a}$, then by definition vector $\mathbf{b}$ is parallel to $\mathbf{a}$ and in the same sense when m is positive but in the opposite sense when m is negative.

Using the symbols with the meanings stated for vector algebra vector equations can be formed and operated on in the same manner as in number algebra.

Exercise 4.

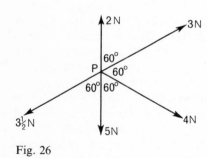

Fig. 26

1. Five coplanar forces 2 N, 3 N, 4 N, 5 N, and $3\frac{1}{2}$ N act at a point P in a body in the directions shown. Find the resultant of the five forces in magnitude and direction. What force would be needed to act at P to produce equilibrium (i.e. a null vector)?

2. Solve the equation

$$3\mathbf{a} = (x+y)\mathbf{a}+(x-y)\mathbf{b}$$

given that $\mathbf{a}$ and $\mathbf{b}$ are non-parallel vectors.

3. Vector $\mathbf{a}$ is a displacement of 2 cm to the NE. Vector $\mathbf{b}$ is a displacement of 3 cm to the SE. Vector $\mathbf{c}$ is a displacement of $3\frac{1}{2}$ cm to the E. Construct a vector diagram to show

(i) $\mathbf{a}+\mathbf{b}-\mathbf{c}$
(ii) $\mathbf{a}+\mathbf{b}-2\mathbf{c}$
(iii) $3\mathbf{a}+2\mathbf{b}-3\mathbf{c}$
(iv) $4\mathbf{a}-2(2\mathbf{a}-\mathbf{b})$
(v) $2\mathbf{a}-\frac{1}{2}(\mathbf{b}-2\mathbf{c})$.

4. Vector $\mathbf{a}$ is a displacement of 3 cm to the NE and vector $\mathbf{b}$ is a displacement of 4 cm in a direction N 30° W. Starting at any

point P find the position Q such that $\mathbf{PQ} = \mathbf{a}$, then find R such that $\mathbf{QR} = \mathbf{b}$. Hence find the vector $\mathbf{c} = \mathbf{a}+\mathbf{b}$ and show it on the vector diagram. Next starting at S, a point on PQ such that $\mathbf{PS} = \mathbf{SQ}$, construct $\mathbf{ST} = \mathbf{d} = \frac{1}{2}\mathbf{a}+\frac{1}{2}\mathbf{b}$.

What is the relation between $\mathbf{c}$ and $\mathbf{d}$? What conclusion do you draw about $\triangle PQR$ and the points S, T? What is this relation called in the geometry of Euclid?

5. Using vectors $\mathbf{a}$ and $\mathbf{b}$ of question 4 show on a vector diagram drawn accurately the vector $\mathbf{a}-\mathbf{b}$. How is the vector $\mathbf{b}-\mathbf{a}$ represented on the same diagram? On another diagram draw the vector $\mathbf{c}-\mathbf{d}$ where $\mathbf{c} = \mathbf{a}$ and $\mathbf{d} = \mathbf{b}$. Which congruency theorem is illustrated by these vector diagrams?

6. Solve the vector equations, given that $\mathbf{a}$, $\mathbf{b}$, and $\mathbf{x}$ are non-parallel vectors:

(i) $3\mathbf{a}+4\mathbf{x} = 8\mathbf{b}$

(ii) $m(\mathbf{a}-\mathbf{b})+\mathbf{x} = \mathbf{a}$

(iii) $\dfrac{s}{s+t}(\mathbf{a}-\mathbf{b})+\mathbf{x} = \mathbf{a}$

7. (i) If $\mathbf{OP} = m\mathbf{OQ}$ and $\mathbf{OR} = \frac{1}{2}\mathbf{OQ}$, what can you say about the points O, P, Q, R?

(ii) If, also, $\mathbf{RO}+\mathbf{OQ} = \frac{1}{4}\mathbf{OP}$, find value of m.

(iii) If the point S lies outside the line OP, deduce a relation between the vectors $\mathbf{SO}$, $\mathbf{SQ}$, and $\mathbf{SP}$.

8. If $\mathbf{P} = (3x+y)\mathbf{a}+(x+y+1)\mathbf{b}$
$\mathbf{Q} = (x-y+4)\mathbf{a}-(y-2x-1)\mathbf{b}$
and $\mathbf{P} = 3\mathbf{Q}$, where vectors $\mathbf{a}$ and $\mathbf{b}$ are non-parallel, find x and y.

9. If O is any point inside a triangle ABC, and L, M, N are the mid-points of the sides AB, BC, CA respectively,

(i) what is $\mathbf{AB}+\mathbf{BC}+\mathbf{CA}$?

(ii) prove that $\mathbf{OB}+\mathbf{OA}+\mathbf{OC} = \mathbf{OL}+\mathbf{OM}+\mathbf{ON}$.

10. If P is any point outside the triangle ABC, and L, M, N are the mid-points of the sides AB, BC, CA respectively, prove that $\overline{PB}+\overline{PA}+\overline{PC} = \overline{PL}+\overline{PM}+\overline{PN}$.

Chapter 3
Vector methods in plane geometry

In the history of mathematics the achievement of Descartes in reducing geometrical problems to algebraic treatment marked a turning point, but it is doubtful if he himself knew just how fundamental a change he had originated. He showed that a mathematical model of geometrical curves could be constructed in purely algebraic terms. The conic sections had previously been treated in a rather laborious geometrical fashion. The treatment he originated was algebraic and from the structure came to be known as co-ordinate geometry. Algebra has often been called the greatest labour-saving device in all mathematics, and Descartes made use of the algebra of his time which we now call *number algebra*—the algebra of points.

During the nineteenth century other algebras were developed. Two of them, vector algebra and matrix algebra, are very closely interlinked, both being offshoots of the work of Hamilton in quaternions and of Grassmann in *Ausdehnungslehre*, that is they originated in the work of these two men as a side product. We shall show later some of the connections between these two algebras when the work in vector algebra has been extended to cover a much wider field, but at this stage we are in a position to show how vector algebra can provide a new approach to many geometrical problems. The main purpose of the next section is not merely to provide a few alternative 'proofs' to some well-known geometrical theorems, but also to show how vectorial methods can provide a new approach in geometry by using directed line segments, or displacement vectors, and the laws of vector algebra on page 25.

Vector geometry

The mathematician needs the help of a real model to understand many abstract concepts. For the concept of a vector quantity we find the help of the geometrical model of a directed line segment indispensable in the early stages. This forms the basis of a new approach to geometry. Many geometrical properties can be deduced from the fundamental definition of a vector quantity and the axioms of vector algebra. The axioms or laws given on page 25 apply to two dimensional vectors at this stage of our treatment; until we have extended our new algebra into three dimensions we shall therefore limit our geometrical treatment to plane geometry.

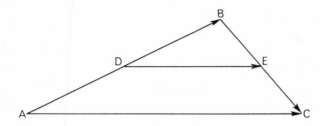

Fig. 27

Treating the triangle ABC vectorially gives

AB+**BC** = **AC**.   (i)

If D and E are the mid-points of AB and BC then by previous work

DB = $\frac{1}{2}$**AB** and **BE** = $\frac{1}{2}$**BC**.

Using scalar multiplication on the vector equation (i) above

$\Rightarrow\frac{1}{2}$(**AB**+**BC**) = $\frac{1}{2}$**AC**

$\Rightarrow\frac{1}{2}$**AB**+$\frac{1}{2}$**BC** = $\frac{1}{2}$**AC**

$\Rightarrow$**DB**+**BE** = $\frac{1}{2}$**AC**.

But from triangle DBE it is seen that **DB**+**BE** = **DE**

$\Rightarrow$**DE** = $\frac{1}{2}$**AC**.

This single vector equation yields two pieces of information: (i) DE is parallel to AC (ii) length DE is half the length AC.
    The result is of course the well known Mid Point Theorem.

It was shown earlier that as long as the symbols are given their correct vectorial meanings, then vector equations can be operated by the usual methods of number algebra. When applied to geometrical problems, methods using vector equations yield important results. Consider now a parallelogram.

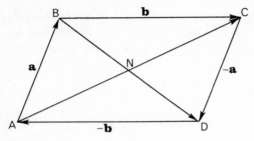

Fig. 28

From definitions if **AB** = **a** then **CD** = −**a** and if **BC** = **b** then
DA = −**b**. If the diagonals intersect at N then vector **BN** can be
written as *n***BD** because vector **BN** has the same direction and
sense as vector **BD**, similarly **AN** can be written as *m***AC**. In the
triangle ABN, **AB** + **BN** = **AN**

$$\Rightarrow \mathbf{a} + \mathbf{BN} = \mathbf{AN}$$

$$\Rightarrow \mathbf{a} = \mathbf{AN} - \mathbf{BN}$$

$$\Rightarrow \mathbf{a} = m\mathbf{AC} - n\mathbf{BD}$$

$$\Rightarrow \mathbf{a} = m(\mathbf{a} + \mathbf{b}) - n[\mathbf{b} + (-\mathbf{a})]$$

$$\Rightarrow \mathbf{a} = (m + n)\mathbf{a} + (m - n)\mathbf{b}.$$

This vector equation can only be true *vectorially* if $m + n = 1$ and
$m - n = 0$.
(*Note*: *two* scalar equations follow from *one* vector equation)
$m = n = \frac{1}{2}$.
This means that **BN** = $\frac{1}{2}$**BD** and **AN** = $\frac{1}{2}$**AC** so N is the midpoint
of BD and AC. In other words the diagonals of the parallelogram
ABCD bisect each other.

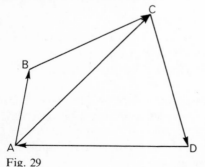

Fig. 29

30

On page 29 the triangle was treated as the sum of three vectors; in the quadrilateral ABCD above it is clear that the sides can be regarded as the vectorial sum also of the four vectors **AB**, **BC**, **CD** and **DA** from the treatment on page 29. We can treat the vectors in pairs giving

$$\mathbf{AB} + \mathbf{BC} + \mathbf{CD} + \mathbf{DA} = (\mathbf{AB} + \mathbf{BC}) + (\mathbf{CD} + \mathbf{DA})$$

But $\mathbf{AB} + \mathbf{BC} = \mathbf{AC}$ and $\mathbf{CD} + \mathbf{DA} = \mathbf{CA} = -\mathbf{AC}$

$$\Rightarrow \mathbf{AB} + \mathbf{BC} + \mathbf{CD} + \mathbf{DA} = \mathbf{AC} + (-\mathbf{AC})$$

$$= \mathbf{0}.$$

Similarly by extending the process we can show that for any polygon the vectorial sum of the sides taken in order is zero.

(This result appears in mechanics as the Polygon of Forces which says that if a system of forces *acting at a point* are in equilibrium they can be represented in magnitude and direction by the sides of a closed polygon taken in order.)

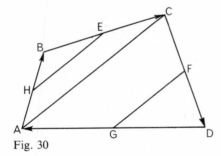

Fig. 30

In the quadrilateral ABCD above, H, E, F and G are the mid-points of the sides AB, BC, CD and DA respectively. We have just shown that

$$\mathbf{AB} + \mathbf{BC} + \mathbf{CD} + \mathbf{DA} = \mathbf{0}.$$

Using scalar multiplication and pairing the vectors gives

$$\tfrac{1}{2}(\mathbf{AB} + \mathbf{BC}) + \tfrac{1}{2}(\mathbf{CD} + \mathbf{DA}) = \mathbf{0}$$

$$\Rightarrow \tfrac{1}{2}(\mathbf{AB} + \mathbf{BC}) = -\tfrac{1}{2}(\mathbf{CD} + \mathbf{DA})$$

$$\Rightarrow \tfrac{1}{2}\mathbf{AB} + \tfrac{1}{2}\mathbf{BC} = -\tfrac{1}{2}\mathbf{CD} - \tfrac{1}{2}\mathbf{DA}$$

$$\mathbf{HB} + \mathbf{BE} = -\mathbf{FD} - \mathbf{DG}$$

$$\mathbf{HE} = -(\mathbf{FD} + \mathbf{DG}) = -\mathbf{FG} = \mathbf{GF}.$$

But if the vectors **HE** and **GF** are equal vectors, they are equal in magnitude and parallel and if the lines HE and GF are equal and parallel to each other then the figure HEFG is a parallelogram. This result means that the lines joining the mid-points of the sides of any quadrilateral form a parallelogram.

To show that the medians of a triangle trisect each other.

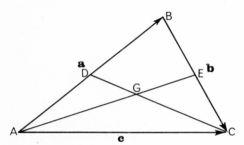

Fig. 31

DC and EA are two of the medians intersecting at G. The position of G is to be determined.

CG can be expressed as k**CD** and **AG** as m**AE**. It will be convenient to let **AB** = **a**, **BC** = **b** and **AC** = **c**. We know that **a** + **b** = **c**.

$$\mathbf{CG} = k\mathbf{CD} = k(-\mathbf{c} + \tfrac{1}{2}\mathbf{a}) = k(-\mathbf{a} - \mathbf{b} + \tfrac{1}{2}\mathbf{a}) = k(-\tfrac{1}{2}\mathbf{a} - \mathbf{b})$$

$$\mathbf{AG} = m\mathbf{AE} = m(\mathbf{c} - \tfrac{1}{2}\mathbf{b}) = m(\mathbf{a} + \mathbf{b} - \tfrac{1}{2}\mathbf{b}) = m(\mathbf{a} + \tfrac{1}{2}\mathbf{b}).$$

But in the triangle AGC, **AG** = **AC** + **CG**

$$m(\mathbf{a} + \tfrac{1}{2}\mathbf{b}) = \mathbf{c} + k(-\tfrac{1}{2}\mathbf{a} - \mathbf{b}) = (\mathbf{a} + \mathbf{b}) - \tfrac{1}{2}k\mathbf{a} - k\mathbf{b}$$

$$\Rightarrow (m + \tfrac{1}{2}k - 1)\mathbf{a} = (1 - k - \tfrac{1}{2}m)\mathbf{b}.$$

Since **a** and **b** are unequal vectors this equation can only be true if

$$(m + \tfrac{1}{2}k - 1) = 0 \quad \text{and} \quad (1 - k - \tfrac{1}{2}m) = 0$$

$$\Rightarrow m + \tfrac{1}{2}k = 1 \quad \text{and} \quad \tfrac{1}{2}m + k = 1$$

$$\Rightarrow m = \tfrac{2}{3} \quad \text{and} \quad k = \tfrac{2}{3}.$$

This shows that G trisects AE and DC, and similarly we can show that G also trisects the median from B. The three medians are therefore concurrent and trisected by the point G.

Position vectors

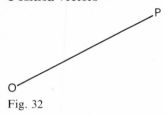

Fig. 32

The concept of the vector enables us to approach geometry in a new fashion and the use of position vectors opens up new ideas.

Consider a variable point P in a plane and let O be any point in the same plane chosen for convenience as origin.

The position of P in relation to the fixed point O is fully determined by the length, direction and sense of the line OP, i.e. the position of P is fully determined by the vector **OP**. Under these conditions **OP** is called the position vector of P. (This idea can be extended to three dimensions by considering any convenient origin O and a variable point P in space.)

The use of position vectors does *not* require the use of coordinate axes, the use of the word origin for O simply means a fixed reference point in the plane chosen for convenience. If however a pair of coordinate axes is imposed on the plane so that their origin coincides with the arbitrarily chosen origin, then the components of the position vector (in the sense given on page 21) would be the x and y coordinates.

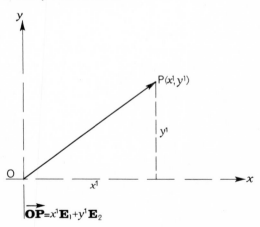

$$\overrightarrow{\mathbf{OP}} = x^1\mathbf{E}_1 + y^1\mathbf{E}_2$$

Fig. 33

If A and B are two points in a plane there is only one straight line passing through them. Let O be any origin chosen in a convenient position, then **OA** and **OB** are the position vectors of A and B with respect to O. Hence if the position vectors **OA** and **OB** are given, then the direction of the straight line joining A and B is determined.

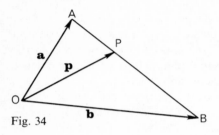

Fig. 34

Any variable point P on the straight line AB will also be determined by its position vector **OP** and the relation between the position vectors **OA**, **OP** and **OB** will be a characteristic of the given line AB.

Since the point P is on the line AB then vector **AP** is parallel to vector **AB**. (Here is a case where we regard a vector as being parallel to itself.) The vector **AP** can be written as k**AB**. If P is the midpoint of AB then **AP** = $\frac{1}{2}$**AB**. If P divides AB in the ratio of $m:n$ then **AP** = $m/(m+n)$**AB**. These two last statements are true because vectors **AP** and **AB** differ only in magnitude not in direction or sense.

To find the position vector of the midpoint of a straight line

Suppose that in Figure 34 AB is the given straight line and P is its midpoint. O is the origin chosen in any convenient position in the plane. Then **OA** and **OB** are the position vectors of the ends of the given straight line. **OP** is the position vector of the midpoint of AB and is to be determined in terms of **OA** and **OB**. **OA** = **a** and **OB** = **b**. Let **OP** = **p**.

p = **a**+**AP** so that **AP** = **p**−**a**

AB = **b**−**a**.

But

AP = $\frac{1}{2}$**AB**

p−**a** = $\frac{1}{2}$(**b**−**a**)

p = **a**+$\frac{1}{2}$(**b**−**a**)

= $\frac{1}{2}$(**a**+**b**).

34

This can be written

$$OP = \tfrac{1}{2}(OA + OB).$$

This is a special case of the problem where the point P divides a given line in a given ratio, say $m:n$, which will now be dealt with.

To find the position vector of the point P which divides the straight line AB in a given ratio.

In Figure 34, let P divide AB in the ratio $m:n$, then $AP = m/(m+n)AB$. The origin O is chosen at any convenient point and $OA = a$, $OP = p$ and $OB = b$.

$$AP = p - a$$

$$AB = b - a$$

and

$$AP = \frac{m}{m+n}AB$$

$$p - a = \frac{m}{m+n}(b - a)$$

$$p = a + \frac{m}{m+n}(b - a)$$

$$OP = \frac{na + mb}{m+n}.$$

If P is the midpoint of AB then $m = n$ and the above relation becomes $OP = (a+b)/2$, which agrees with the result previously obtained for the same condition.

To find the equation of a straight line in *vector form*

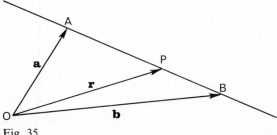

Fig. 35

Suppose the straight line whose vector equation is required, passes through the given points A and B whose position vectors are **a** and **b** with reference to the origin O.

OA = **a** and **OB** = **b**.

Now if P is any point in the line and **r** is its position vector
From Figure 35

OA + **AP** = **OP**.

We have shown that vector equations can be operated like the equations of number algebra and so

AP = **OP** − **OA**.

$\quad$ = **r** − **a**

also

OA + **AB** = **OB**

$\quad \Rightarrow$ **AB** = **OB** − **OA**

$\quad\quad$ = **b** − **a**.

But since **AP** and **AB** are collinear then the variable length AP can be expressed as a scalar multiple of the fixed length AB.
In vector form
AP = t **AB** where t is a scalar variable which can be $+ve$, or $-ve$ if P lies on BA produced.

$\Rightarrow$ **r** − **a** = $t(\mathbf{b} - \mathbf{a})$

$\quad \Rightarrow$ **r** = **a** + $t(\mathbf{b} - \mathbf{a})$

$\quad \Rightarrow$ **r** = **a** + $t\mathbf{b} - t\mathbf{a}$

$\quad\quad$ = $(1 - t)\mathbf{a} + t\mathbf{b}$.

This can be written

$(1 - t)\mathbf{a} + t\mathbf{b} - \mathbf{r} = \mathbf{0}$.

This is the equation of the straight line with reference to the arbitrary or convenient origin O.

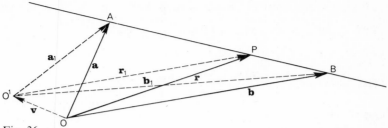

Fig. 36

If another origin O' is now chosen then using the suffix notation to show the new position vectors $\mathbf{a}_1, \mathbf{b}_1, \mathbf{r}_1$ from this origin where $\mathbf{O'B} = \mathbf{b}_1$.

$$(1-t)\mathbf{a}+t\mathbf{b}-\mathbf{r} = \mathbf{0} \ldots \text{ for the origin O}$$

and

$$\mathbf{a} = \mathbf{v}+\mathbf{a}_1 \qquad \mathbf{b} = \mathbf{v}+\mathbf{b}_1 \qquad \mathbf{r} = \mathbf{v}+\mathbf{r}_1$$

$$\Rightarrow (1-t)(\mathbf{v}+\mathbf{a}_1)+t(\mathbf{v}+\mathbf{b}_1)-(\mathbf{v}+\mathbf{r}_1) = \mathbf{0}$$

$$\mathbf{v}(1-t+t-1)+(1-t)\mathbf{a}_1+t\mathbf{b}_1-\mathbf{r}_1 = \mathbf{0}$$

$$\mathbf{v}0+(1-t)\mathbf{a}_1+t\mathbf{b}_1-\mathbf{r}_1 = \mathbf{0}$$

$$\Rightarrow (1-t)\mathbf{a}_1+t\mathbf{b}_1-\mathbf{r}_1 = \mathbf{0}.$$

Therefore the form of the equation is independent of the origin chosen, the constants alter for different origins but the variable t is unaffected.

Vectorial treatment of Centroids

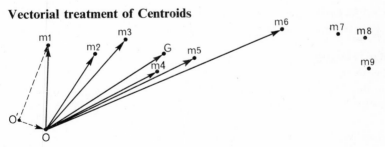

Fig. 37

If $m_1, m_2, m_3, m_4, m_5, m_6, \ldots m_n$ are n points in a plane then for any origin O the position vectors of the n points are $\mathbf{Om}_1, \mathbf{Om}_2, \mathbf{Om}_3, \mathbf{Om}_4, \ldots \mathbf{Om}_n$.

If G is some point in the plane such that

$$n\mathbf{OG} = \mathbf{Om}_1 + \mathbf{Om}_2 + \mathbf{Om}_3 + \mathbf{Om}_4 + \cdots + \mathbf{Om}_n$$

then

$$\mathbf{OG} = \frac{\mathbf{Om}_1 + \mathbf{Om}_2 + \mathbf{Om}_4 + \cdots + \mathbf{Om}_n}{n} = \frac{\Sigma \mathbf{Om}_r}{n}$$

and the point G is defined as the CENTROID of the n points.

We can show that the position of the point G does *not* depend on the position of the arbitrary origin chosen by taking another origin O', then if G' is the centroid for this origin

$$\mathbf{O'm}_1 = \mathbf{O'O} + \mathbf{Om}_1, \; \mathbf{O'm}_2 = \mathbf{O'O} + \mathbf{Om}_2 \text{ etc., and by definition}$$

$$\mathbf{O'G'} = \frac{\mathbf{O'm}_1 + \mathbf{O'm}_2 + \mathbf{O'm}_3 + \cdots + \mathbf{O'm}_n}{n}$$

$$= \frac{\mathbf{O'O} + \mathbf{Om}_1 + \mathbf{O'O} + \mathbf{Om}_2 + \mathbf{O'O} + \mathbf{Om}_3 + \cdots + \mathbf{O'O} + \mathbf{Om}_n}{n}$$

$$= \frac{n\mathbf{O'O} + \mathbf{Om}_1 + \mathbf{Om}_2 + \mathbf{Om}_3 + \cdots + \mathbf{Om}_n}{n}$$

$$= \mathbf{O'O} + \mathbf{OG}$$

$$= \mathbf{O'G}.$$

Thus $\mathbf{O'G'} = \mathbf{O'G}$ and this can only be true if G' coincides with G. This means that although another origin has been used the centroid has remained fixed. The centroid based on the original definition is a fixed point in the system.

Mass centroid

If with the points $m_1, m_2, m_3, m_4, \ldots m_r, \ldots m_n$ we associate a scalar number M_r which measures the mass at these points then the centroid of the masses is defined as before:

$$(M_1 + M_2 + M_3 + \cdots + M_n)\mathbf{OG} =$$

$$M_1\mathbf{Om}_1 + M_2\mathbf{Om}_2 + M_3\mathbf{Om}_3 + \cdots + M_n\mathbf{Om}_n$$

$$\mathbf{OG} = \frac{M_1\mathbf{Om}_1 + M_2\mathbf{Om}_2 + M_3\mathbf{Om}_3 + \cdots + M_n\mathbf{Om}_n}{M_1 + M_2 + M_3 + M_4 + \cdots + M_n} = \frac{\Sigma M_r\mathbf{Om}_r}{\Sigma M_r}$$

The position vector **OG** so defined gives the position of the mass centroid.

With the points $m_1, m_2, m_3, m_4, \ldots m_n$ we could associate scalar numbers to represent area, weight, volume etc., and in a similar manner find the area centroid, weight centroid (usually called the centre of gravity), or the volume centroid.

Non-collinear vectors

Definition

Two vectors are said to be non-collinear vectors when they are *not* parallel to the same line.

Being free vectors we can arrange them so that their initial points coincide and they are then coplanar. Any other vector in *this* plane can be taken to have its initial point starting at their point of intersection and can then be expressed as a function of the other two uniquely, i.e. in only *one* possible way.

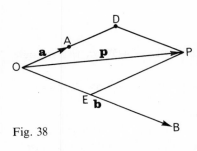

Fig. 38

Let **OA** = **a** and **OB** = **b** be two such non-collinear vectors chosen quite arbitrarily and **OP** = **p** a third vector in their plane. If **ODPE** is a parallelogram **p** = **OP** = **OD** + **OE** = $l\mathbf{a} + m\mathbf{b}$ where l and m are scalars, positive or negative depending on the directions. **a** and **b** are called the BASE VECTORS and $l\mathbf{a}$ and $m\mathbf{b}$ the COMPONENT VECTORS. Now suppose we could express $\mathbf{p} = s\mathbf{a} + t\mathbf{b}$. This would mean that

$$l\mathbf{a} + m\mathbf{b} = s\mathbf{a} + t\mathbf{b} \Rightarrow (l-s)\mathbf{a} = (t-m)\mathbf{b}$$

But this can only be true if **a** is parallel to **b** which is not possible by the given conditions so

$$l - s = 0 \quad \text{and} \quad t - m = 0$$

is the only alternative

$$\Rightarrow s = l \quad \text{and} \quad t = m.$$

Hence

p is uniquely given by $l\mathbf{a} + m\mathbf{b}$.

We have now shown that in two dimensions any vector can be expressed uniquely in terms of two base vectors which are not collinear and is then said to be linearly dependent on them. We say that two non-collinear vectors are sufficient *to span* two-dimensional space, and any third vector in that space can be expressed in terms of the two base vectors. We shall show later that three non-coplanar vectors *span* three-dimensional space, i.e. any 3-vector can be uniquely expressed in terms of the three base-vectors, i.e. four 3-vectors in three-dimensional space must be linearly dependent.

Any three vectors in a plane must be linearly dependent vectors.

Exercise 4a

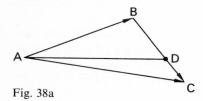

Fig. 38a

1. Referring to Fig. 38a: given that $3\mathbf{BD} = 4\mathbf{DC}$ in the triangle ABC, simplify $3\mathbf{AB} + 4\mathbf{AC}$.

2. In the triangle ABC of Fig. 38a, D cuts the side AC internally in the ratio of $\lambda:\mu$.
Prove that $\mu\overline{\mathbf{AB}} + \lambda\overline{\mathbf{AC}} = (\mu + \lambda)\,\overline{\mathbf{AD}}$.

3. O, B, C are the points $(0, 0)$, $(3, 4)$, and $12, -5)$. If force of 20 N acts along OB, and a force of 26 N acts along OC, find the magnitude and direction of the resultant force.

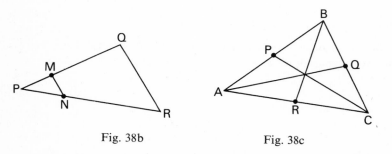

Fig. 38b   Fig. 38c

4. In $\triangle$PQR, M and N cuts the sides PQ and PR internally in the same ratio $m:n$.
Show that MN is parallel to QR.

5. In the triangle ABC, P, Q, and R are the mid-points of sides AB, BC, and CA respectively. If vectors **a**, **b**, and **c**, represent the vectors $\overline{AB}$, $\overline{BC}$, and $\overline{CA}$ Express $\overline{AQ}$, $\overline{BR}$, and $\overline{CP}$ in terms of **a**, **b**, and **c**, hence find the sum of the three median vectors.

6. The position vectors of the vertices A, B, C of the triangle ABC taking the circumcentre O as origin, are denoted by **a**, **b**, and **c**.

(i) Where is the point P whose position vector **p** is given by $\mathbf{p} = \mathbf{a} + \mathbf{c}$?

(ii) the point Q_1, whose position vector is given by $\mathbf{q}_1 = (\mathbf{a} + \mathbf{c}) + \mathbf{b}$?

(iii) the point R whose position vector is given by $\mathbf{r} = \mathbf{a} + \mathbf{b}$ and Q_2 given by $\mathbf{q}_2 = (\mathbf{a} + \mathbf{b}) + \mathbf{c}$?

(iv) the point T whose position vector is given by $\mathbf{t} = \mathbf{b} + \mathbf{c}$ and Q_3 given by $\mathbf{q}_3 = (\mathbf{b} + \mathbf{c}) + \mathbf{a}$?

Show that the points Q_1, Q_2, Q_3 are coincident. Calling the coincident point H, show that it is the orthocentre. Show that the centroid of the points A, B, and C is on the line joining O and H.

7.

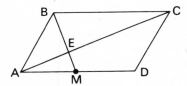

Fig. 38d

M is the mid-point of side AD of the parallelogram ABCD. BM cuts the diagonal AC at E. By what ratio is AC cut by BM at E?

8. Prove that the joins of mid-points of adjacent sides of a skew (concave) quadrilateral form a parallelogram using the property that the vector sums of the sides of a closed polygon is the zero vector.

9. Show that the vector sum of the medians of a triangle is zero.

10. Referring to Fig. 38a, a force of magnitude 3AB acts along AB and another force of magnitude 2AC acts along AC. Give the magnitude and direction of the resultant force.

If any three coplanar vectors can be expressed in a linear relation such as $l\mathbf{a} + m\mathbf{b} + n\mathbf{c} = 0$ (where l, m and n are not *all* equal to zero) the three vectors are defined as *linearly dependent vectors*. This subject will be dealt with later.

Having shown something of the power of vector algebra in geometrical work, it ought to be made clear that vector methods are not advocated for all such problems, only where they illuminate the work more closely.

We now come to another aspect of vectors. The engineer and the scientist know very clearly what they mean by a vector quantity and we have shown how such quantities can be represented completely by displacement vectors or line-segments. The mathematician has found that he can represent vectors in a plane by a pair of ordered numbers, or vectors in space by a triple of ordered numbers. Many people have come to talk of such vectors as 2-vectors or 3-vectors for obvious reasons. One of the great extensions of this idea has been the concept of mathematical 'quantities' which need four or more numbers to specify them completely, a concept which has grown rapidly with the wide extension of computer facilities to linear programming. It is difficult to visualise a 'space' of four or more dimensions, and it is impossible to construct a physical model of such spaces, but the mathematician not only conceives of such spaces, but uses them, calling them hyperspaces since they are not real in the physical sense.

A 'mathematical quantity' which requires *n* ordered numbers to specify it completely is called a 'vector of *n* dimensions'. The vector algebra which concerns itself with more than three dimensions is called abstract vector algebra of *n* dimensions. We shall limit ourselves to two and three dimensions, in the early part of this book.

Consider the vector **AB** = **p**. It is the result of compounding or

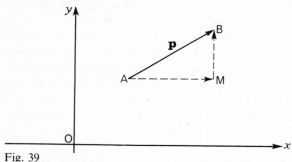

Fig. 39

adding vectorially the displacement **AM** horizontally and **MB** vertically.

If AM is say 3 units long and MB is 2 units long. Then vector **AM** is 3 times the *unit vector* horizontally, and vector **MB** is 2 times the *unit vector* vertically.

So the two *numbers* 3 and 2 specify the number of *unit vectors* horizontally and vertically which will give the vector **AB**.

$\mathbf{p} = 3\hat{\mathbf{x}} + 2\hat{\mathbf{y}}$ where $\hat{\mathbf{x}}$ and $\hat{\mathbf{y}}$ are the unit vectors horizontally and vertically. $\hat{\mathbf{x}}$ and $\hat{\mathbf{y}}$ have been called CAP *x* and CAP *y*.

We can regard 3 and 2 as *coefficients* necessary to specify the vector **P** in terms of the *unit vectors* already referred to.

The algebra of detached coefficients is of course MATRIX ALGEBRA.* The detached coefficients can be expressed as a single column matrix $\begin{bmatrix} 3 \\ 2 \end{bmatrix}$ with the usual notation to enclose the matrix.

Hence the column matrix $\begin{bmatrix} 3 \\ 2 \end{bmatrix}$ specifies the vector **AB** completely.

Sometimes the *column matrix* is called a *column vector*.

We started with a definite vector **AB** whose horizontal component is 3 unit vectors, whose vertical component is 2 unit vectors.

Vector **a** can be represented by $\begin{bmatrix} 4 \\ 3 \end{bmatrix}$

Vector **b** can be represented by $\begin{bmatrix} -2 \\ 3 \end{bmatrix}$

Vector **c** can be represented by $\begin{bmatrix} 4 \\ -1 \end{bmatrix}$

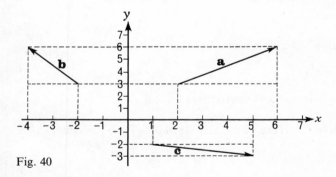

Fig. 40

Using $\hat{\mathbf{x}}$ and $\hat{\mathbf{y}}$ to represent as before the UNIT VECTORS horizontally and vertically the following vector equations are true.

$\mathbf{a} = 4\hat{\mathbf{x}} + 3\hat{\mathbf{y}}$

* See A. E. Coulson, *Introduction to Matrices*, Longmans, 1965.

3

$$\mathbf{b} = -2\hat{\mathbf{x}} + 3\hat{\mathbf{y}}$$

$$\mathbf{c} = 4\hat{\mathbf{x}} - \hat{\mathbf{y}}$$

Remembering that in vector algebra the LHS and RHS of the equations represent vectors, + means the *vectorial sum,* and the − means the vectorial difference already defined, i.e. $-2\hat{\mathbf{x}}$ means a horizontal vector of magnitude two units to the left

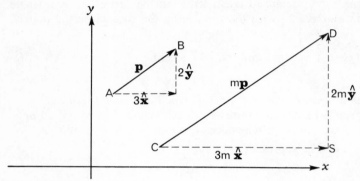

Fig. 41

Suppose the vector **p** has components $3\hat{\mathbf{x}}$ and $2\hat{\mathbf{y}}$ then

$$\mathbf{p} = 3\hat{\mathbf{x}} + 2\hat{\mathbf{y}}$$

and we can use the column vector $\begin{bmatrix} 3 \\ 2 \end{bmatrix}$ to represent **p**.

If **p** is multiplied by the scalar number *m* then

$$m\mathbf{p} = m(3\hat{\mathbf{x}} + 2\hat{\mathbf{y}})$$

$$= 3m\hat{\mathbf{x}} + 2m\hat{\mathbf{y}} \text{ by the Distribution Law.}$$

So the column vector representing *m*p is $\begin{bmatrix} 3m \\ 2m \end{bmatrix}$ which can be shown $m\begin{bmatrix} 3 \\ 2 \end{bmatrix}$.

In the figure the triangle CDS has all its sides *m* times those of triangle ABT and it is obvious by similar triangle properties that CD(*m*p) is parallel to AB(**p**).

The vector represented by $m \begin{bmatrix} 3 \\ 2 \end{bmatrix}$ is therefore parallel to the vector represented by $\begin{bmatrix} 3 \\ 2 \end{bmatrix}$ and has m times the magnitude.

Exercise 5.
1. On squared paper mark the points P ($x = 3$, $y = 4$) and Q ($x = 5$, $y = 3$). If $\hat{x}$ and $\hat{y}$ are the unit vectors along the x-axis and y-axis respectively and **p** and **q** are the position vectors of the points P and Q, express **p** and **q** in terms of the unit vectors $\hat{x}$ and $\hat{y}$. Express the vector **PQ** in terms of **p** and **q** and hence express **PQ** in terms of $\hat{x}$ and $\hat{y}$.
2. Find the position vector of M the mid-point of PQ as given in question 1.
3. Find the position vector of the centroid of the points A(3, 4), B(5, 3), and C(7, 8).
4. Use the position vectors of the points A(2, 3), B(4, 4) and C(6, 5) to find **AB** and **BC** and hence show that A, B and C are collinear.
5. If G is the centroid of the triangle ABC prove that

GA + GB + GC = 0

6. If O is any point in the plane of triangle ABC show that **OA + OB + OC** = 3**OG**, where G is the centroid of triangle ABC.
7. Show that the lines joining the midpoints of opposite sides of a tetrahedron are concurrent and bisect each other.

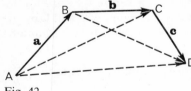

Fig. 42

8. In the figure **AB**, **BC** and **CD** represent vectors **a**, **b** and **c**. What vectors are represented by **AC** and **BD**? Give **AD** as the sum of two vectors and state your conclusions.

9. In △ABC the midpoints of BC, CA and AB are D, E and F respectively. O is any point, prove that

OA + OB + OC = **OF + OD + OE**

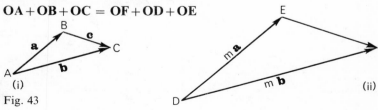

Fig. 43

(i)

(ii)

10. From figure (i) express **c** in terms of **a** and **b** and from figure (ii) express **EF** in terms of **a**, **b** and the scalar *m*. What is the relation between |**c**| and |**EF**|? What conclusions do you draw about the lengths and directions of the sides of the two triangles? What is the name given to such triangles?

11. Masses of $2m$, $3m$ and $2m$ are placed at the points $(3, 5)$ $(5, 3)$ and $(7, 8)$ respectively. What is the position vector of their centre of mass? Give the coordinates of the centroid.

12. Given points A $(1, 6)$ and B $(5, 3)$ find the position vector of the point D which divides the line AB in the ratio of $2:1$.

13. If the vertices A, B, and C of a triangle ABC have position vectors **a**, **b**, and **c** respectively, from some origin 0 in the plane of ABC,

   (i) show that the centroid of the points A, B, and C is given by the position vector $\dfrac{\mathbf{a}+\mathbf{b}+\mathbf{c}}{3}$.

   (ii) If equal masses M are placed at A, B, and C, where will the mass centroid be situated?

14. Masses of M, $3m$, $2m$, and $4m$ are placed at the vertices A, B, C, and D respectively of the parallelogram ABCD. Taking 0, the point of intersection of the diagonals AC and BD, as origin, and calling the position vectors of A and B, **a** and **b** respectively, find:

   (i) the position vector of G the mass centroid;

   (ii) the position of the point G.

Addition of column vectors

If $\mathbf{a} = m\hat{\mathbf{x}} + n\hat{\mathbf{y}}$ and $\mathbf{b} = l\hat{\mathbf{x}} + p\hat{\mathbf{y}}$, then $\mathbf{a}+\mathbf{b} = (m+l)\hat{\mathbf{x}} + (n+p)\hat{\mathbf{y}}$ by the distribution law so we can represent the sum of **a** and **b** by the column vector $\begin{bmatrix} (m+l) \\ (n+p) \end{bmatrix}$ simply by the addition of corresponding elements which agrees with the process of matrix addition

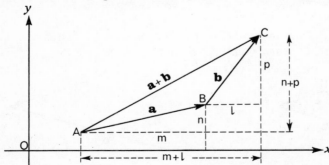

Fig. 44

The diagram shows the addition of **a** and **b** by the usual vector method of addition. **AC** is the sum of **a** and **b**. The vector **AC** has horizontal component $(m+l)\hat{\mathbf{x}}$ and vertical component $(n+p)\hat{\mathbf{y}}$. This shows that the addition of two column vectors gives precisely the same end result as the addition of the corresponding vectors by the triangle method.

Once more we see that a purely geometrical process can be represented algebraically by the use of column vectors (or column matrices).

Subtraction of column vectors
We now demonstrate that the subtraction of column vectors gives the same result as the geometrical vector process outlined earlier.

Consider the same two vectors

$$\mathbf{a} = m\hat{\mathbf{x}}+n\hat{\mathbf{y}}$$

$$\mathbf{b} = l\hat{\mathbf{x}}+p\hat{\mathbf{y}}$$

$$\mathbf{a}-\mathbf{b} = m\hat{\mathbf{x}}+n\hat{\mathbf{y}}-l\hat{\mathbf{x}}-p\hat{\mathbf{y}}$$

$$= (m-l)\hat{\mathbf{x}}+(n-p)\hat{\mathbf{y}}$$

Hence the column vector $\begin{bmatrix}(m-l)\\(n-p)\end{bmatrix}$ represents $\mathbf{a}-\mathbf{b}$

In the diagram the horizontal distance from A to D is $(m-l)$ units, the vertical distance is downwards (with the lengths given i.e. $-ve$ and equal to $(n-p)$).

In the figure $\mathbf{a}-\mathbf{b} = \mathbf{AD}$.

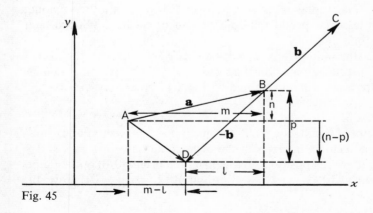

Fig. 45

So **AD** is the difference of **a** and **b**, it also has the components arrived at by algebraic subtraction of the elements of the column vectors. Hence the geometrical process and algebraic processes are equivalent.

Magnitude of a 2-vector

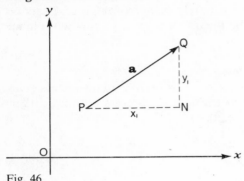

Fig. 46

The displacement vector **PQ** represents the vector quantity **a**.

$$\mathbf{PQ} = \mathbf{PN} + \mathbf{NQ}$$

$\mathbf{PN} = x_1\hat{\mathbf{x}}$ where $\hat{\mathbf{x}}$ is the unit vector in the base direction Ox. Similarly $\mathbf{NQ} = y_1\hat{\mathbf{y}}$ where $\hat{\mathbf{y}}$ is the unit vector in the base direction Oy.

PN is x_1 units long, NQ is y_1 units long

$$|\mathbf{a}| = |\mathbf{PQ}| = \sqrt{\mathrm{PN}^2 + \mathrm{NG}^2}$$
$$= \sqrt{x_1^2 + y_1^2}.$$

(The magnitude of **a** can be established in another manner using the scalar product, a concept which will be developed later in this book.)

Since the unit vectors $\hat{\mathbf{x}}$ and $\hat{\mathbf{y}}$ are at right angles to each other they constitute a pair of orthogonal base vectors spanning a plane, they are more usually denoted by **i** and **j** respectively.

Exercise 6.

1. Vector $\mathbf{a} = 6\hat{\mathbf{x}} + 8\hat{\mathbf{y}}$. Find the magnitude of vector **a**. Express the unit vector $\hat{\mathbf{a}}$ in terms of the orthogonal base vectors $\hat{\mathbf{x}}$ and $\hat{\mathbf{y}}$.
2. What are the position vectors from the origin $(0, 0)$ of the points determined by A $(x = 3,\ y = 4)$ and B $(x = 6,\ y = 7)$? Give the vector **AB**

(i) in terms of the unit vectors $\hat{\mathbf{x}}$ and $\hat{\mathbf{y}}$

(ii) as a column vector.

3. The position vectors of points A and B are $\mathbf{a} = 2\hat{\mathbf{x}} + 4\hat{\mathbf{y}}$ and $\mathbf{b} = 6\hat{\mathbf{x}} + 7\hat{\mathbf{y}}$. The point P has the position vector $\mathbf{p} = 8\hat{\mathbf{x}} + 9\hat{\mathbf{y}}$. Find the vectors **AB** and **AP**. Are the points A, B and P collinear?

4. The vertices of the $\triangle ABC$ are $A(2, 3)$, $B(4, 5)$, $C(3, 7)$. Give **AB**, **BC** and **AC** in terms of the unit vectors $\hat{\mathbf{x}}$ and $\hat{\mathbf{y}}$ and then give **AB** + **BC**. Find the centroid of the three points A, B and C, giving its position vector.

5. Vector **AB** has magnitude 5 units and is parallel to vector $\mathbf{p} = 2\hat{\mathbf{x}} + 3\hat{\mathbf{y}}$. Express **AB** in terms of its component vectors $\hat{\mathbf{x}}$ and $\hat{\mathbf{y}}$.

6. Vector $\mathbf{r} = 3\hat{\mathbf{x}} + 2\hat{\mathbf{y}}$. Find vectors **a**, **b** and **c** such that $\mathbf{a} = -\mathbf{r}$, $\mathbf{b} = 6\mathbf{r}$ and $\mathbf{c} = -k\mathbf{r}$.

7. If $\mathbf{a} = 2\hat{\mathbf{x}} + 3\hat{\mathbf{y}}$ and $\mathbf{b} = 3\hat{\mathbf{x}} - 2\hat{\mathbf{y}}$, find the unit vectors parallel to $\mathbf{a} + \mathbf{b}$, $\mathbf{a} - \mathbf{b}$, and $2\mathbf{a} + 3\mathbf{b}$.

Linearly independent vectors

If **a** and **b** are non-parallel vectors in a plane then $m\mathbf{a} + n\mathbf{b} = \mathbf{0}$ only if $m = n = 0$.

For suppose that m is not equal to zero then

$$\mathbf{a} = \frac{-n\mathbf{b}}{m}$$

but this implies that **a** is parallel to **b** which is contrary to the given condition. It must follow then that $n = 0$ and so $m = 0$ also.

Two vectors which cannot be expressed in the relation $m\mathbf{a} + n\mathbf{b} = \mathbf{0}$ unless $m = n = 0$ are said to be linearly independent. If m and n can have non-zero values in the relation $m\mathbf{a} + n\mathbf{b} = 0$, then it follows that vectors **a** and **b** must be parallel vectors and the one can always be expressed in terms of the other; in this case the two vectors are said to be linearly dependent.

If **a**, **b** and **c** are three non-parallel vectors in a plane, then any one vector, say **a** can always be expressed in terms of the other two vectors **b** and **c**.

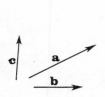

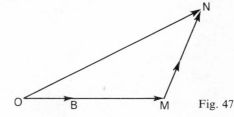

Fig. 47

If **OB** represents vector **b** let **OM** represent $m\mathbf{b}$. From the terminal point M let **MN** be drawn to represent $n\mathbf{c}$, then **ON** is the resultant **d** and $\mathbf{d} = m\mathbf{b} + n\mathbf{c}$. So for all non-zero values of **b** and **c** the resultant **d** can be made parallel to *any* direction in the plane by using suitable values of the scalars m and n as long as both m and n are not zero. Hence **d** can be made parallel to **a**. With suitable values of m and n then,

$\mathbf{d} = l\mathbf{a}$ where l is a scalar.

But $\mathbf{d} = m\mathbf{b} + n\mathbf{c}$

$l\mathbf{a} = m\mathbf{b} + n\mathbf{c}$

or,

$l\mathbf{a} - m\mathbf{b} - n\mathbf{c} = \mathbf{0}$.

Since l, m and n can take positive or negative values this relation is more usually written

$l\mathbf{a} + m\mathbf{b} + n\mathbf{c} = \mathbf{0}$

where l, m and n are not all zero.

Since any one vector can be expressed in terms of the other two, they are said to be linearly dependent. Two non-parallel vectors in a plane have been shown to be linearly independent, now we have shown that three non-parallel vectors in a plane are linearly dependent. This means that two non-parallel vectors are sufficient to *span* two-dimensional space.

Example

If $\mathbf{a} = \begin{bmatrix} 3 \\ 4 \end{bmatrix}$ $\mathbf{b} = \begin{bmatrix} 5 \\ 6 \end{bmatrix}$ $\mathbf{c} = \begin{bmatrix} 7 \\ 8 \end{bmatrix}$ express **c** in terms of **a** and **b**.

Let $\mathbf{c} = m\mathbf{a} + n\mathbf{b}$

$$\begin{bmatrix} 7 \\ 8 \end{bmatrix} = m\begin{bmatrix} 3 \\ 4 \end{bmatrix} + n\begin{bmatrix} 5 \\ 6 \end{bmatrix}$$

$$= \begin{bmatrix} 3m \\ 4m \end{bmatrix} + \begin{bmatrix} 5n \\ 6n \end{bmatrix}$$

$$= \begin{bmatrix} 3m + 5n \\ 4m + 6n \end{bmatrix}.$$

Since we are dealing with 2-vectors we can extract two pieces of information from the vector equation that the column vector $\begin{bmatrix} 7 \\ 8 \end{bmatrix}$

is equal to the column vector $\begin{bmatrix} 3m+5n \\ 4m+6n \end{bmatrix}$

$3m+5n = 7$

$4m+6n = 8$

and $m = -1$, $n = 2$ satisfies this pair of simultaneous equations. We conclude therefore that

$\mathbf{c} = -1\mathbf{a}+2\mathbf{b}.$

In a similar manner we could express $\mathbf{a}$ or $\mathbf{b}$ in terms of the other two vectors. The three vectors are linearly dependent.

In the previous example the vectors were given in column vector form. We now work an example in component form.

Example

If $\mathbf{s} = 2\hat{\mathbf{x}}+3\hat{\mathbf{y}}$ and $\mathbf{t} = 3\hat{\mathbf{x}}+4\hat{\mathbf{y}}$ find a vector $\mathbf{w}$ such that $\mathbf{w} = 4\mathbf{s}-2\mathbf{t}$.

$\mathbf{w} = 4(2\hat{\mathbf{x}}+3\hat{\mathbf{y}})-2(3\hat{\mathbf{x}}+4\hat{\mathbf{y}})$

$= 8\hat{\mathbf{x}}-6\hat{\mathbf{x}}+12\hat{\mathbf{y}}-8\hat{\mathbf{y}}$

$= 2\hat{\mathbf{x}}+4\hat{\mathbf{y}}$

Exercise 7.

1. If $\mathbf{a} = \begin{bmatrix} 1 \\ 2 \end{bmatrix}$ $\mathbf{b} = \begin{bmatrix} 5 \\ 6 \end{bmatrix}$ $\mathbf{c} = \begin{bmatrix} 7 \\ 8 \end{bmatrix}$ express $\mathbf{a}$ in terms of $\mathbf{b}$ and $\mathbf{c}$.

2. If $\mathbf{p} = 4\frac{1}{2}\hat{\mathbf{x}}+6\hat{\mathbf{y}}$ and $\mathbf{q} = 10\hat{\mathbf{x}}+12\hat{\mathbf{y}}$, find a vector $\mathbf{r}$ such that $\mathbf{r} = 3\mathbf{p}-2\mathbf{q}$.

3. If $m\mathbf{s}+n\mathbf{t} = 3\mathbf{w}$ find the scalars m and n such that

$\mathbf{s} = \begin{bmatrix} 2 \\ 3 \end{bmatrix}$ $\mathbf{t} = \begin{bmatrix} 3 \\ 5 \end{bmatrix}$ and $\mathbf{w} = \begin{bmatrix} 4 \\ 8 \end{bmatrix}$

3*

4. The vectors **a**, **b** and **c** have the magnitudes and directions shown in the following diagram

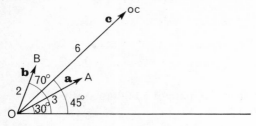

Fig. 48

Show by a geometrical diagram how the vector **c** can be expressed as linearly dependent on the vectors **a** and **b**. (*Hint*: use the directions of **a** and **b** as base directions and **a** and **b** as base vectors.)

5. In triangle OBC denote vector **OB** by **b**, vector **OC** by **c**.
  (i) What is the unit vector along OB?
 (ii) What is the unit vector along OC?

(iii) If $\mathbf{OX} = m\left(\dfrac{\mathbf{b}}{|\mathbf{b}|} + \dfrac{\mathbf{c}}{|\mathbf{c}|}\right)$, what can you say about **OX**?

6. OABC is a rhombus. If **OA** = **a** and **OC** = **c**, give **AB** and **BC** in terms of **a** and **c**; give also **AC** and **OB** in terms of **a** and **c**. What is the relation between $(\mathbf{c} - \mathbf{a})$ and $(\mathbf{c} + \mathbf{a})$?

7. The position vectors **p** and **r** of the points P and R are given by $\mathbf{p} = \begin{pmatrix} 2 \\ 8 \end{pmatrix}$ and $\mathbf{r} = \begin{pmatrix} 6\frac{1}{2} \\ 2 \end{pmatrix}$. Find the position vector **q** of the point Q which cuts PR internally in ratio of $2:1$. The position vector **s** of the point S is $\mathbf{s} = \begin{pmatrix} \frac{1}{2} \\ 10 \end{pmatrix}$. What can you say about P, Q, R, S?

Chapter 4
Multiplication of vectors

Development of the concept of multiplication of two vectors

Different kinds of vector quantities can be represented by displacement vectors. The operations carried out with these displacement vectors represent also the operations that take place with all these different kinds of vector quantities. So far the operations with which we have dealt have been the addition of the *same kind of vectors* and the multiplication of a vector quantity by a scalar; in each of our operations it must be emphasised that only the same kind of vector quantities can be added, the addition of a force vector to a velocity vector is not defined and has no meaning. At the end of the eighteenth century and the beginning of the nineteenth century, physicists, engineers and mathematicians were making great discoveries about vector quantities but the very word VECTOR had not yet been invented by Sir William Hamilton. The engineers like James Watt who introduced the term 'horsepower' had found that by multiplying the magnitude of the force overcome, by the distance through which the point of application of the force had moved gave a scalar quantity which they called WORK DONE. In 1820, Ampère made his famous discovery that a conductor carrying an electric current in a magnetic field was acted upon by a mechanical force at right angles to the geometrical plane containing the electric current and the magnetic field. In 1831, after six years of experiment, Michael Faraday discovered the converse effect, namely that a conductor moving at right angles to a magnetic field has an electric current induced in it and in a direction at *right angles* to the geometrical plane containing the magnetic field and the motion. These were great discoveries but it was not until 1843 that Sir W. R. Hamilton presented to the Royal Irish Academy a mathematical treatment, which he published in 1853. In 1844, H. G. Grassmann published a treatment called the *Ausdehnungslehre* (the Theory of Extensions) which included most of Hamilton's ideas and extended them much further, but from quite independent work. For the first time mathematicians recognised in these two works vector quantities as such, and it was Hamilton who first used the word VECTOR in its present meaning (derived from the Latin verb *vehere*, to carry) but he used the term in connection with the quaternions which he had invented; to him a vector was part of a quaternion which was composed of a SCALAR and a VECTOR part. For some years following 1853 mathematicians tried enthusiastically to employ quaternions in science and engineering, but they made little headway, finding they were useful only in a very limited field. Concurrently with these developments, brilliant mathematicians were making remarkable advances in the applications of algebra to new fields: the algebra of classes (Boolean algebra), matrix algebra, Group theory and Complex Numbers. It is only in our own lifetime that the importance of these nineteenth-century discoveries is being realised, a process hastened stupendously by the development of the COMPUTER since 1947. The classical algebra of the pre-nineteenth-century period is correctly called *number algebra* to distinguish it from the various modern varieties. In this traditional algebra, a symbol such as 'x' always represented a number, but in modern algebras a symbol could represent a SET of objects, a CLASS of statements, a Matrix, a Group, or a Complex Number etc. Towards the end of the nineteenth century, Professor J. W. Gibbs of Yale and O. Heaviside in England, working independently, developed an algebra of vectors which was largely adapted from the work of Hamilton on quaternions and Grassmann's *Ausdehnungslehre*. Heaviside first suggested the use of bold type as the algebraic symbol to represent vector quantities and the normal Roman type to represent scalars. In typing or writing we cannot imitate bold type so the vector symbols are shown by underlining. An Italian school of mathematicians also developed an algebra of vectors, and some of the symbols they used are still in use, but less frequently now.

The work of Watt, Ampère and Faraday had shown that although the addition of two different kinds of vectors had no physical meaning, they could interact in such a way that the process of multiplication is involved.

When the two different vectors were acting in the same direction a scalar quantity was produced, but when they acted at right angles they produced *another vector quantity*.

In both these physical processes, 'multiplication' is involved, but it is a new sort of combination of different kinds of vector quantities which we call 'multiplication' because of the similarity of its results to those of multiplication in number algebra. So in vector algebra two kinds of multiplications are encountered and to be meaningful the mathematical processes used must produce results which are in accordance with known physical processes. In the next section we shall try to illustrate how algebraic multiplication and vector multiplications are related, but first we must show the discoveries of Ampère and Faraday in relation to vectors of electrical and magnetic nature.

In 1819 Professor Oersted of Copenhagen discovered that a conductor carrying an electric current is surrounded by a magnetic field; at any point in this field a magnetic force acted on a magnetic pole, and this is a vector field. Oersted showed that a wire carrying an electric current placed over and parallel to a magnetic compass needle caused a deflection of the needle; reversing the current caused the reversal of the deflection; he showed that this was due to the magnetic field round the conductor which is illustrated in Fig. 49; the direction of the lines of magnetic force show the

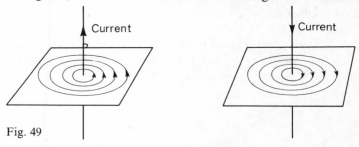

Fig. 49

direction in which a free north pole would tend to move if placed in that field, if the electric current is constant then the magnetic field remains in a steady state.

Then in 1820 Ampère made his famous discovery of the force acting on a current-carrying conductor placed in a magnetic field. This can be shown quite simply by passing an electric current along a wire placed between the poles of a magnet as in the diagram.

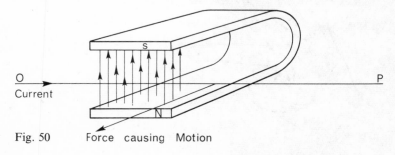

Fig. 50   Force causing Motion

Ampère found that if a current flows in the direction from O to P, then the wire is deflected outwards in a direction at right angles to the magnetic field and to the current. This is shown diagrammatically below:

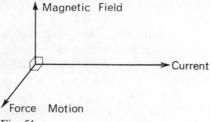

Fig. 51

If the magnetic force vector and the current vector are in the plane of this page then the force vector causing the deflection of the wire would be acting outwards and perpendicular to the plane of the paper.

Had the current been passed from P to O, i.e. reversed in direction, then the wire would have been deflected inwards towards the centre of the magnet; this is shown below:

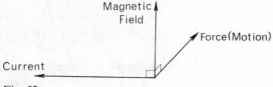

Fig. 52

Ampère's discovery can be remembered by the application of FLEMING'S LEFT-HAND RULE:

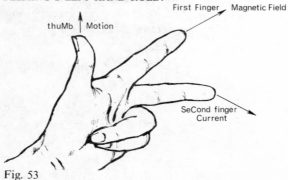

Fig. 53

If the letters in capitals are emphasised they form an easy aid to memory and the three fingers so held indicate three directions mutually perpendicular to each other. Whenever the action of an electric current through a conductor acts on a magnetic field to produce motion, then Fleming's left-hand rule is applicable.

In 1831 Michael Faraday discovered the converse effect namely that if a conductor is moved across, i.e. at right angles to, a magnetic field, then an electric current is induced in the conductor. This can be demonstrated in the laboratory by using the arrangement of diagram (54).

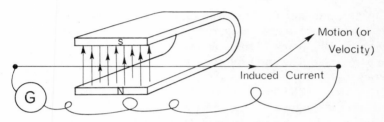

Fig. 54

In Faraday's experiment the velocity vector of the conductor across the magnetic field vector produced an electric current in the conductor as shown. We can show this diagrammatically:

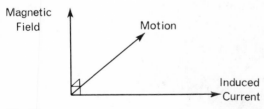

Fig. 55

In this diagram, if the magnetic field vector and the velocity vector are in a plane perpendicular to the plane of this paper then the induced current lies in the plane of the paper and at right angles to the other two vectors. Had the motion been reversed then the induced current would have been reversed also:

57

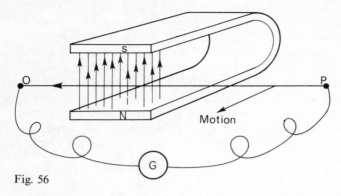

Fig. 56

If the conductor OP is moved rapidly away from the centre of the horseshoe magnet, i.e. at right angles to the direction of the magnetic field, then a current induced flows in the direction shown from P to O. The moment the motion ceases then the current ceases.

Referring to the diagram it is seen once more that the three vectors, the magnetic force, the velocity and electric current are mutually at right angles to each other. In this case we can also consider that the electric current vector is at right angles to the plane containing the other two vectors.

When motion of a conductor in a magnetic field produces an electric current then Fleming's RIGHT-HAND rule is applicable. This rule is similar to the left hand rule, the thuMb represents the Motion, the First Finger represents the Field of Force and the seCond finger the Current but the right hand is used instead of the left:

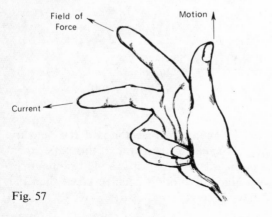

Fig. 57

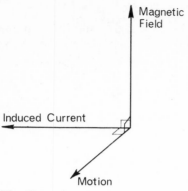

Fig. 58

These discoveries of Ampère and Faraday were discoveries of the fundamental properties of vector quantities and can be repeated in any laboratory with the aid of a horseshoe magnet, a length of copper wire and a sensitive galvanometer (for most purposes a milliammeter serves excellently). Ampère showed also that the magnitude of the third vector produced was equal to the *product* of the magnitudes of the two vectors which generated it. Faraday found the same result. Some years earlier it had been shown that the magnitude of the scalar quantity *work done* was also equal to the product of the magnitudes of the *force overcome* and the displacement vector in the same direction.

So by 1831, although vectors were not defined until 1844, mathematicians had been shown by the engineers and physicists that the product of two vectors in the same direction was a scalar quantity and that the product of two vectors at right angles to each other was a vector quantity in a direction at right angles to the plane containing the first two vectors. By 1844, Hamilton and Grassmann had produced a mathematical treatment of vectors, in the case of Hamilton limited to vectors in three dimensions, but Grassmann's treatment was general and not limited to three dimensions only. However, the methods used by Hamilton and Grassmann did not become very acceptable to scientists or engineers and in 1881, Professor Willard Gibbs adapted those parts of the methods of both Hamilton and Grassmann which were acceptable and laid the foundations of vector algebra; Heaviside working independently in England had produced a similar treatment.

59

Multiplication of vectors*

Consider two vectors **a** and **b** of different kinds, and in the plane containing them choose two base axes at right angles. Let vector **a** make an angle of ϕ_1 with the x-axis and vector **b** an angle ϕ_2 also with the x-axis:

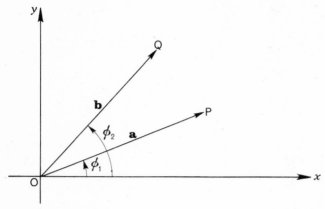

Fig. 59

The vector **a** can be expressed in terms of its component vectors along the x and y axes

$$\mathbf{a} = |\mathbf{a}| \cos \phi_1 \hat{\mathbf{x}}_1 + |\mathbf{a}| \sin \phi_1 \hat{\mathbf{y}}_1$$

Similarly vector **b** can be expressed but the unit vectors must be distinguished from those used for vector **a** since they are different kinds of vectors

$$\mathbf{b} = |\mathbf{b}| \cos \phi_2 \hat{\mathbf{x}}_2 + |\mathbf{b}| \sin \phi_2 \hat{\mathbf{y}}_2$$

Assuming that we can multiply vector **a** by vector **b** we have

$$\mathbf{ab} = (|\mathbf{a}| \cos \phi_1 \hat{\mathbf{x}}_1 + |\mathbf{a}| \sin \phi_1 \hat{\mathbf{y}}_1)(|\mathbf{b}| \cos \phi_2 \hat{\mathbf{x}}_2 + |\mathbf{b}| \sin \phi_2 \hat{\mathbf{y}}_2)$$

$$= |\mathbf{a}| |\mathbf{b}| \cos \phi_1 \cos \phi_2 \hat{\mathbf{x}}_1 \hat{\mathbf{x}}_2 + |\mathbf{a}| |\mathbf{b}| \sin \phi_1 \sin \phi_2 \hat{\mathbf{y}}_1 \hat{\mathbf{y}}_2$$

$$+ |\mathbf{a}| |\mathbf{b}| \cos \phi_1 \sin \phi_2 \hat{\mathbf{x}}_1 \hat{\mathbf{y}}_2 + |\mathbf{a}| |\mathbf{b}| \sin \phi_1 \cos \phi_2 \hat{\mathbf{y}}_1 \hat{\mathbf{x}}_2$$

Since $\hat{\mathbf{x}}_1$ and $\hat{\mathbf{x}}_2$, $\hat{\mathbf{y}}_1$ and $\hat{\mathbf{y}}_2$ are unit vectors in the *same direction,* the products $\hat{\mathbf{x}}_1 \hat{\mathbf{x}}_2$ and $\hat{\mathbf{y}}_1 \hat{\mathbf{y}}_2$ can be assumed to be *unit* scalar quantities from the results of experimental work. But since $\hat{\mathbf{x}}_1 \hat{\mathbf{y}}_2$ and $\hat{\mathbf{y}}_1 \hat{\mathbf{x}}_2$ are the products of two vectors which are *right angles,* experimental work shows that in these cases the products are unit

* See: pp. 80, 81 and 82. Mathematical Association, *A second report on the teaching of mechanics in schools.* Bell, 1965.

vectors at right angles to the plane containing the unit vectors $\hat{\mathbf{x}}_1, \hat{\mathbf{x}}_2, \hat{\mathbf{y}}_1$ and $\hat{\mathbf{y}}_2$, but $\hat{\mathbf{y}}_1\hat{\mathbf{x}}_2$ will be opposite in direction to $\hat{\mathbf{x}}_1\hat{\mathbf{y}}_2$ as indicated in the experiments. We can now write for **ab**

$$= (|\mathbf{a}|\,|\mathbf{b}|\cos\phi_1\cos\phi_2 + |\mathbf{a}|\,|\mathbf{b}|\sin\phi_1\sin\phi_2)$$

$$+ (|\mathbf{a}|\,|\mathbf{b}|\cos\phi_1\sin\phi_2\hat{\mathbf{z}} + |\mathbf{a}|\,|\mathbf{b}|\sin\phi_1\cos\phi_2[-\hat{\mathbf{z}}])$$

$$= |\mathbf{a}|\,|\mathbf{b}|(\cos\phi_1\cos\phi_2 + \sin\phi_1\sin\phi_2)$$

$$+ |\mathbf{a}|\,|\mathbf{b}|(\cos\phi_1\sin\phi_2 - \sin\phi_1\cos\phi_2)\hat{\mathbf{z}}$$

$$= |\mathbf{a}|\,|\mathbf{b}|\cos(\phi_2-\phi_1) + |\mathbf{a}|\,|\mathbf{b}|\sin(\phi_2-\phi_1)\hat{\mathbf{z}}$$

But $(\phi_2-\phi_1) =$ angle between **a** and **b**, call it θ, hence

$$\mathbf{ab} = |\mathbf{a}|\,|\mathbf{b}|\cos\theta + |\mathbf{a}|\,|\mathbf{b}|\sin\theta\,\hat{\mathbf{z}}$$

The first part of this product is a *scalar quantity* and the second part is a *vector quantity* at right angles to the plane containing vectors **a** and **b**.

It has become standard practice to distinguish between the scalar part of the product and the vector part, by calling the scalar part the DOT product of the two vectors and the vector part the CROSS product. This is shown

$$\mathbf{a}\,.\,\mathbf{b} = |\mathbf{a}|\,|\mathbf{b}|\cos\theta \;(\theta \text{ is the angle between } \mathbf{a} \text{ and } \mathbf{b})$$

$$\mathbf{a}\times\mathbf{b} = |\mathbf{a}|\,|\mathbf{b}|\sin\theta\hat{\mathbf{u}} \text{ (where } \hat{\mathbf{u}} \text{ is the unit vector perpendicular to the plane containing } \mathbf{a} \text{ and } \mathbf{b})$$

Sometimes the vector product is denoted $\mathbf{a}\wedge\mathbf{b}$ instead of $\mathbf{a}\times\mathbf{b}$ and for some purposes the DOT product is often called the INNER product.

Now in the preceding treatment, at the commencement we multiplied vector **a** by vector **b** in that order, and we considered angles measured in a positive direction from Ox so that the angle θ is also measured in the positive direction *from* vector **a**. If however we multiplied vector **b** by vector **a** still measuring the angles ϕ_1 and ϕ_2 from the direction Ox, then the angle θ_1 would be measured in the negative direction, i.e. $\theta_1 = (\phi_1-\phi_2) = -(\phi_2-\phi_1) = -\theta$ and the sine of a negative angle is opposite in *sign* to the sine of a positive angle $\Rightarrow \sin(-\theta) = -\sin\theta$ and it follows that

$\mathbf{b}\times\mathbf{a} = |\mathbf{b}|\,|\mathbf{a}|\sin(-\theta)\hat{\mathbf{u}}$ Since **a** and **b** are scalars we can write

$\qquad = -|\mathbf{a}|\,|\mathbf{b}|\sin\theta\hat{\mathbf{u}}$ $|\mathbf{b}|\,|\mathbf{a}| = |\mathbf{a}|\,|\mathbf{b}|$

$\qquad = -\mathbf{a}\times\mathbf{b}$

Thus the cross product is a non-commutative multiplication since
b × a = −a × b.

This consideration of the sign of the cross product can be approached another way. In all our mathematics where the three axes of reference are used, Ox, Oy, Oz, it has become standard practice to use a right-handed system of coordinates.

The right-handed system of coordinates

If as in Figure 60 the right hand is put with the palm facing upward, the third and fourth fingers are closed and the second finger is held at right angles to the palm, then the thumb is taken to point in the direction Ox, the first (or index) finger points in the direction Oy and the second finger points in the direction Oz. This is shown in the following diagram:

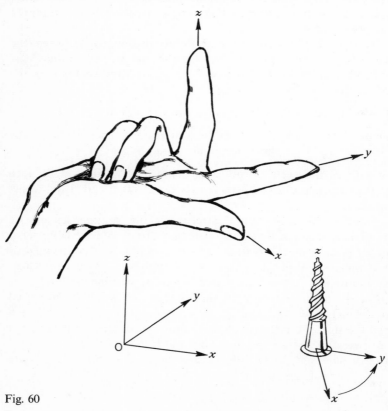

Fig. 60

Referring to diagram 60 if Ox and Oy are intended to be in the plane of the paper then Oz is intended to point vertically *upwards* out of the plane of Ox and Oy. This agrees with experiment for if a vector acting along Ox is multiplied by a vector acting along Oy then the vector (or cross) product would be a vector acting in the direction Oz. It will be remembered that Ampère found that a current vector acting along Ox, in a magnetic field acting along Oy suffered a force vector in the direction Oz. Ampère also found that if the direction of the current was reversed then the force vector acted along Oz in the opposite direction, i.e. in a negative direction to the original force. This is shown in the next diagram.

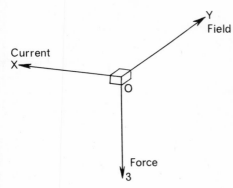

Ox and Oy lie in the plane of the page and Oy points downwards through the *back* of the page.

Fig. 61

But if the thumb of the right hand points in the new direction of the current and the index finger in the same direction of the field this can only be done if the right hand is now held palm downwards, with the second finger pointing downwards. This means that if we are to standardise on a right-handed system of coordinates, multiplication of vector **a** by vector **b** results in a vector having the direction in which a right handed screw would move if turned in the direction from **a** to **b**. Hence once again this means that $\mathbf{a} \times \mathbf{b} = -\mathbf{b} \times \mathbf{a}$. In practice the idea of the movement of the common wood screw is one of the simplest methods of deciding on the direction of the cross product. The next diagrams illustrate this idea. In both cases the vector **c** is at right angles to the shaded plane containing the vectors **a** and **b**.

In coordinate geometry of three dimensions, analytical geometry and in all other branches of mathematics the right handed system

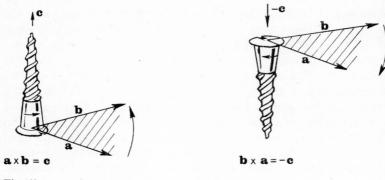

a × **b** = **c**   **b** × **a** = **-c**

Fig. 62

is in general use and the idea of positive and negative directions included to give a system which we now illustrate.

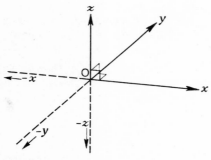

Fig. 63

Chapter 5
Scalar product of two vectors

Multiplication of vectors

In the early part of this book the multiplication of a vector by a scalar was shown geometrically and algebraically; the process was indicated in vector algebra by putting the scalar number in front of the vector symbol. The multiplication of vector **a** by scalar m is shown m**a**, no special sign being used to indicate the process. We now know that the algebraic multiplication of two vectors in component form results in a product of two parts—one part is a scalar and the other part is a vector. The extraction of the scalar part is called scalar multiplication, and is denoted by a dot; the extraction of the vector part is called vector multiplication and is denoted by the × or ∧ sign. Scalar multiplication is often spoken of as DOT multiplication and vector multiplication as CROSS multiplication.

Vector multiplication involves a vector at right angles to the plane of the vectors which are being multiplied. The process takes place in *three dimensions* so vector multiplication will be dealt with after vectors in three dimensions have been studied.

The projection of a vector

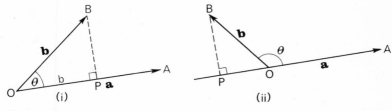

Fig. 64

OB represents vector **b** making an angle θ with the direction of vector **a** represented by **OA**. BP is perpendicular to the direction OA.

The length OP is defined as the PROJECTION of vector **b** on vector **a**. The projection of one vector on another is by definition a SCALAR. (If OP has the direction shown in figure (ii) then the

65

length OP is multiplied by -1.) It is clear that $\mathbf{OB} = \mathbf{OP} + \mathbf{PB}$ and the vectors $\mathbf{OP}$ and $\mathbf{PB}$ are called the component vectors of $\mathbf{OB}$ and the scalar $|\mathbf{OP}|$ is also called the COMPONENT of vector $\mathbf{b}$ in the direction $\mathbf{a}$. Since any vector in the direction of $\mathbf{OA}$ can be represented by $k\hat{\mathbf{a}}$ where k is the magnitude of the vector and $\hat{\mathbf{a}}$ is the unit vector in the direction of $\mathbf{a}$, it follows that the vector $\mathbf{OP}$ of magnitude $|\mathbf{OP}|$ and direction OA can be represented by $|\mathbf{OP}|\hat{\mathbf{a}}$. As OP is the projection of $\mathbf{b}$ on the direction OA then $|\mathbf{OP}| = |\mathbf{b}| \cos \theta$ and $\mathbf{OP} = |\mathbf{b}| \cos \theta \hat{\mathbf{a}}$.

The scalar (or dot) product

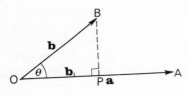

Fig. 65

Definition

If the length of the projection of vector $\mathbf{b}$ on vector $\mathbf{a}$ is denoted by $|\mathbf{b}_1|$, then the SCALAR PRODUCT of the two vectors $\mathbf{a}$ and $\mathbf{b}$, $\mathbf{a} \cdot \mathbf{b} = |\mathbf{a}||\mathbf{b}_1|$.

If θ is the angle between the vectors then $|\mathbf{b}_1| = |\mathbf{b}| \cos \theta$.

Hence $\mathbf{a} \cdot \mathbf{b} = |\mathbf{a}||\mathbf{b}_1|$

$$= |\mathbf{a}||\mathbf{b}| \cos \theta$$

When a force acts on a body and there is a displacement then the point of application of the force moves and *work* is said to be done. Work done is a scalar quantity, it has no sense or direction and is an idea developed at the beginning of the nineteenth century to enable engineers to show how their steam engines could be used instead of horses for pumping water from pits.

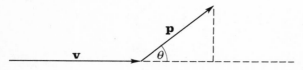

Fig. 66

If the force $\mathbf{p}$ acts at an angle θ to the displacement vector $\mathbf{v}$ then the work done in the direction of the displacement $= |\mathbf{v}||\mathbf{p}| \cos \theta$ because $|\mathbf{p}| \cos \theta$ is the magnitude of the force in the effective direction. This formula for work done is therefore the same as the scalar product, i.e. work done $= \mathbf{v} \cdot \mathbf{p}$, but whereas *work done* only applies to the particular vectors **p** and **v**, the scalar product applies to all vectors.

If two vectors are perpendicular then $\cos 90° = 0$ and the scalar product must be zero—this can be used as a test for perpendicularity.

If two vectors are equal then

$\mathbf{a} \cdot \mathbf{a} = |\mathbf{a}| \, |\mathbf{a}| \cos 0°$ and because $\cos \theta = 1$ since $\theta = 0°$ in this case

$\qquad = |\mathbf{a}| \, |\mathbf{a}| \cdot 1$

$\mathbf{a} \cdot \mathbf{a} = \mathbf{a}^2$ by definition $= |\mathbf{a}|^2$

Consider again the scalar product **a . b**

$\mathbf{a} \cdot \mathbf{b} = |\mathbf{a}| \, |\mathbf{b}| \cos \theta \ldots \ldots$ These are all scalar quantities

$\Rightarrow \mathbf{a} \cdot \mathbf{b} = |\mathbf{b}| \, |\mathbf{a}| \cos \theta$

$\qquad = \mathbf{b} \cdot \mathbf{a}$ $\Rightarrow$ The scalar product is commutative.

Notice that $n\mathbf{a} \cdot \mathbf{b} = |n\mathbf{a}| \, |\mathbf{b}| \cos \theta$

$\qquad\qquad = |\mathbf{a}| \, |n\mathbf{b}| \cos \theta$

$\qquad\qquad = \mathbf{a} \cdot n\mathbf{b} = n(\mathbf{a} \cdot \mathbf{b})$

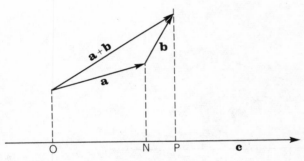

Fig. 67

If **a**, **b** and **c** are three coplanar vectors (2-vectors) then by definition the scalar product of vector $(\mathbf{a}+\mathbf{b})$ and **c**, is $|\mathbf{OP}| \, |\mathbf{c}|$ but

$|\mathbf{OP}| = |\mathbf{ON}| + |\mathbf{NP}|$

$\Rightarrow |\mathbf{c}| \, |\mathbf{OP}| = |\mathbf{c}|[|\mathbf{ON}| + |\mathbf{NP}|]$ since $|\mathbf{ON}|$ is the projection of **a** on **c**

$\qquad\qquad = |\mathbf{c}| \, |\mathbf{ON}| + |\mathbf{c}| \, |\mathbf{NP}|$ and $|\mathbf{NP}|$ is the projection of **b** on **c**

$\mathbf{c} \cdot (\mathbf{a}+\mathbf{b}) = |\mathbf{c}| \, |\mathbf{a}_1| + |\mathbf{c}| \, |\mathbf{b}_1|$

$\qquad\qquad = \mathbf{c} \cdot \mathbf{a} + \mathbf{c} \cdot \mathbf{b}$

i.e. the scalar product is distributive.

By extension, if $\mathbf{c} = \mathbf{l}+\mathbf{m}$

$(\mathbf{l}+\mathbf{m})(\mathbf{a}+\mathbf{b}) = (\mathbf{l}+\mathbf{m}) \cdot \mathbf{a} + (\mathbf{l}+\mathbf{m}) \cdot \mathbf{b}$

$\qquad\qquad = \mathbf{l} \cdot \mathbf{a} + \mathbf{m} \cdot \mathbf{a} + \mathbf{l} \cdot \mathbf{b} + \mathbf{m} \cdot \mathbf{b}$

67

We have now shown that the scalar product is commutative and distributive.

The inner product and the scalar product

If vector **a** is given in row vector form $[x_1, y_1]$ i.e. a row matrix, and vector **b** is also given in column vector form $\begin{bmatrix} x_2 \\ y_2 \end{bmatrix}$ then, in matrix algebra, multiplication of two such matrices gives the inner product $(x_1 x_2 + y_1 y_2)$.

We shall now show that the inner product so defined is equivalent to the scalar product of the two vectors as previously defined.

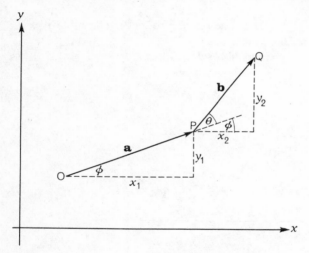

Fig. 68

OP and **PQ** represents the coplanar vectors **a** and **b** of components x_1, y_1 and x_2, y_2.

In the diagram $x_1 = |\mathbf{a}| \cos \phi$, $\qquad y_1 = |\mathbf{a}| \sin \phi$

$$x_2 = |\mathbf{b}| \cos (\theta + \phi), \qquad y_2 = |\mathbf{b}| \sin (\theta + \phi)$$

$$\Rightarrow x_1 x_2 + y_1 y_2 = |\mathbf{a}| \cos \phi \, |\mathbf{b}| \cos (\theta + \phi) + |\mathbf{a}| \sin \theta \, |\mathbf{b}| \sin (\theta + \phi)$$

$$= |\mathbf{a}| |\mathbf{b}| \cos (\theta + \phi) \cos \phi + |\mathbf{a}| |\mathbf{b}| \sin (\theta + \phi) \sin \phi$$

$$= |\mathbf{a}| |\mathbf{b}| \{ \cos (\theta + \phi) \cos \phi + \sin (\theta + \phi) \sin \phi \}$$

$$= |\mathbf{a}| |\mathbf{b}| \{ \cos (\overline{\theta + \phi - \phi}) \}$$

$$= |\mathbf{a}| |\mathbf{b}| \cos \theta$$

but this is the scalar product as defined on page 64. In matrix algebra the product $(x_1 x_2 + y_1 y_2)$ is also called the scalar product (sometimes the inner product).

We have now shown the ISOMORPHISM between 2-vectors and ordered couples of real numbers.

Laws for scalar Product

On page 25 the laws of vector algebra, established for the addition of vectors, were summarised. We can now add laws of vector algebra for scalar Product.

1. Scalar Product is commutative

   $\mathbf{a} \cdot \mathbf{b} = \mathbf{b} \cdot \mathbf{a}$

2. $\mathbf{c} \cdot (\mathbf{a} + \mathbf{b}) = \mathbf{c} \cdot \mathbf{a} + \mathbf{c} \cdot \mathbf{b}$ Distributive Law

3. $(\mathbf{c} + \mathbf{d}) \cdot (\mathbf{a} + \mathbf{b}) = (\mathbf{a} + \mathbf{b})(\mathbf{c} + \mathbf{d})$

   $$= \mathbf{a} \cdot \mathbf{c} + \mathbf{b} \cdot \mathbf{c} + \mathbf{a} \cdot \mathbf{d} + \mathbf{b} \cdot \mathbf{d}$$

4. $n\mathbf{a} \cdot \mathbf{b} = \mathbf{a} \cdot n\mathbf{b} = n(\mathbf{a} \cdot \mathbf{b})$

5. If $\mathbf{a} \cdot \mathbf{b} = 0$ and $\mathbf{a}$ and $\mathbf{b}$ are not null vectors, then $\mathbf{a}$ and $\mathbf{b}$ are perpendicular to each other.

   The scalar (or dot) product has many applications in geometry and we shall now illustrate its value as an alternative method of establishing some well known metrical results.

   In using the result $\mathbf{a} \cdot \mathbf{b} = |\mathbf{a}||\mathbf{b}| \cos \theta$ it must be clear that θ will be limited to positive angles and $0 < \theta < \pi$.

   The Laws of Vector Algebra for scalar products correspond exactly with the Laws of Number Algebra, if we remember the special meanings we give to the symbols.

   We can use the scalar (or dot) product to deduce a number of well known elementary theorems.

Theorem of Pythagoras

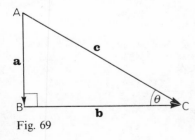

Fig. 69

In the right angled triangle vector $\mathbf{a}$ is perpendicular to vector $\mathbf{b}$ hence

$\mathbf{a} \cdot \mathbf{b} = |\mathbf{a}||\mathbf{b}| \cos \theta = 0$

(since $\cos 90° = 0$).

By definition of vector addition $\mathbf{c} = \mathbf{a} + \mathbf{b}$, using the scalar

69

products of equal vectors on themselves

$\Rightarrow \mathbf{c} . \mathbf{c} = (\mathbf{a}+\mathbf{b}) . (\mathbf{a}+\mathbf{b})$

$\Rightarrow |\mathbf{c}|^2 = \mathbf{a} . \mathbf{a} + \mathbf{a} . \mathbf{b} + \mathbf{b} . \mathbf{a} + \mathbf{b} . \mathbf{b}$

$\Rightarrow |\mathbf{c}|^2 = |\mathbf{a}|^2 + 0 + 0 + |\mathbf{b}|^2$

$\Rightarrow AC^2 = AB^2 + BC^2$

If $\mathbf{c}$ is the unit vector $|\mathbf{a}| = \sin\theta$, $|\mathbf{b}| = \cos\theta$ from earlier definition, i.e.

$$\cos^2\theta + \sin^2\theta = 1$$

from the earlier definition of $\cos\theta$ and $\sin\theta$.

Cosine rule for a triangle ABC

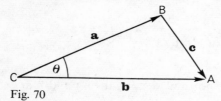

Fig. 70

Since $\mathbf{b} = \mathbf{a}+\mathbf{c}$

$\quad\quad \mathbf{c} = \mathbf{b}-\mathbf{a}$

$\quad \mathbf{c} . \mathbf{c} = (\mathbf{b}-\mathbf{a}) . (\mathbf{b}-\mathbf{a})$

$\quad\quad = \mathbf{b} . \mathbf{b} - \mathbf{a} . \mathbf{b} - \mathbf{b} . \mathbf{a} + \mathbf{a} . \mathbf{a}$

$\quad\quad = \mathbf{b} . \mathbf{b} + \mathbf{a} . \mathbf{a} - 2\mathbf{a} . \mathbf{b}$

$\quad |\mathbf{c}|^2 = |\mathbf{b}|^2 + |\mathbf{a}|^2 - 2\,|\mathbf{a}||\mathbf{b}|\cos\theta$

$\quad AB^2 = CA^2 + CB^2 - 2CB . CA . \cos\theta$

Apollonius' Theorem

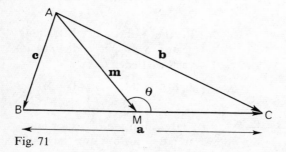

Fig. 71

70

In △ABM

$$c = m - \frac{a}{2}$$

$$\Rightarrow c \cdot c = \left(m - \frac{a}{2}\right) \cdot \left(m - \frac{a}{2}\right)$$

$$= m \cdot m + \frac{a}{2} \cdot \frac{a}{2} - \frac{2a}{2} \cdot m$$

$$\Rightarrow |c|^2 = |m|^2 + \frac{|a|^2}{4} - |a| \, |m| \cos \theta \qquad (1)$$

In △AMC

$$b = m + \frac{a}{2}$$

$$\Rightarrow b \cdot b = \left(m + \frac{a}{2}\right)\left(m + \frac{a}{2}\right)$$

$$= m \cdot m + \frac{a}{2} \cdot \frac{a}{2} + \frac{2 \cdot a \cdot m}{2}$$

$$|b|^2 = |m|^2 + \frac{|a|^2}{4} + |a| \, |m| \cos \theta \qquad (2)$$

Adding (1) and (2)

$$|b|^2 + |c|^2 = 2|m|^2 + \frac{|a|^2}{2}$$

$$\Rightarrow AC^2 + AB^2 = 2AM^2 + \frac{BC^2}{2}$$

Exercise 8.

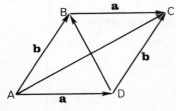

Fig. 72

1. Show that the diagonals of a rhombus are perpendicular to each other.

Using the figure shown express **DB** and **AC** in terms of **a** and **b**, then use the scalar product of **AC . DB**.

2.

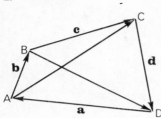

Fig. 73

In the quadrilateral ABCD prove that $\mathbf{AC.BD} = \mathbf{AB.CD} + \mathbf{BC.AD}$. (*Hint*: use the vectors $\mathbf{a}$, $\mathbf{b}$, $\mathbf{c}$ and $\mathbf{d}$ as shown. Note that $\mathbf{AD} = -\mathbf{a}$.)

3. If the quadrilateral ABCD of Qu. 2 is cyclic, show that the result reduces to $\mathbf{AB.CD} + \mathbf{BC.AD} = \mathbf{AC.BD}$. (*Hint:* draw BH to meet AC in H so that $\widehat{ABH} = \widehat{CBD}$, hence $\triangle ABH$ and $\triangle CBD$ are equiangular as ABCD is cyclic.)

4. Use the scalar product to prove that the perpendicular bisectors of the sides of a triangle are concurrent.

5. Two concurrent forces represented by $\mathbf{OP}$ and $\mathbf{OQ}$ make an angle of θ with each other. If $\mathbf{OR}$ represents the resultant of $\mathbf{OP}$ and $\mathbf{OQ}$, use the scalar product to show that

$$OR^2 = OP^2 + 2OP \cdot OQ \cos \theta$$

6. Prove that the sum of the squares of the diagonals is equal to the sum of the squares of the sides.

7. AB is the diameter of a circle of centre O. P is any point on the circumference. Use the scalar product $\mathbf{AP.BP}$ to prove that angle APB is a right-angle.

8. Prove that if two circles intersect, the line joining their centres is perpendicular to the line joining their points of intersection.

9. Find the value of m such that the vectors $\mathbf{a} = 2\hat{\mathbf{x}} + 3\hat{\mathbf{y}}$ and $\mathbf{b} = m\hat{\mathbf{x}} + 2\hat{\mathbf{y}}$ are perpendicular to each other.

10. Find the angle between vectors $\mathbf{p}$ and $\mathbf{q}$ given that $\mathbf{p} = \begin{pmatrix} 3 \\ 4 \end{pmatrix}$, $\mathbf{q} = \begin{pmatrix} 12 \\ -5 \end{pmatrix}$.

Chapter 6
Vectors in three dimensions (3-vectors)

Our work so far has been confined to vectors in two dimensions but since so many applications of vectors in engineering, science and electrical technology require a three-dimensional treatment, this must now be introduced.

The addition and subtraction of three-dimensional vectors follows similar treatment for 2-vectors with algebraic results which appear to be identical in form and structure.

Algebra is a BINARY operation, i.e. *two* quantities are involved in the operations defined; so the processes of addition, subtraction and multiplication are binary operations (subtraction being defined so that it also becomes a process of addition).

In order to deal with three-dimensional space, three base directions (not coplanar) must be defined and it has been found convenient to use a system of *orthogonal* axes which are *right-handed.*

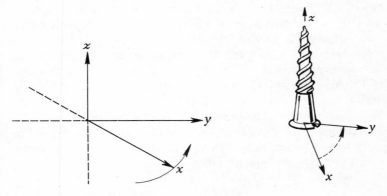

Fig. 74

The three axes shown are called a right-handed system because if we imagine a right-handed wood screw being held so that it is pointed along the direction Oz then on being turned from Ox round towards Oy it would travel in the direction of Oz.

The dotted portions show the *negative* directions of the axes.

73

This convention will be used throughout whenever the need to use coordinate axes arises.

If two vectors are equal they have the same magnitude and the same direction (direction here meaning *orientation in space and sense*), the point of application is not important and often *not meaningful* so that if we wish to add two 3-vectors **a** and **b** we proceed to define addition in a similar manner to our earlier definition.

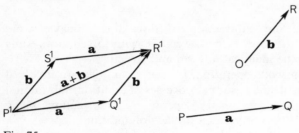

Fig. 75

If the displacement **PQ** represents vector quantity **a** and displacement **OR** represents vector quantity **b**, then by localizing the vectors at **P'Q'** and **Q'R'** the addition is defined as the displacement **P'R'**. A pair of concurrent straight lines define a plane hence the sum of **a** and **b** is coplanar with them.

Inspection will show that P'S' drawn equal and parallel to Q'R' represents **b** and S'R' being the side of the parallelogram P'Q'R'S' opposite to P'Q' must represent **a** also in magnitude and direction.

Hence $\mathbf{a}+\mathbf{b} = \mathbf{b}+\mathbf{a}$. (Commutative law of addition.)

The definition of a negative vector on pages 10 and 25 applies equally to a 3-vector, i.e. $\mathbf{a}+(-\mathbf{a}) = 0$. Hence it follows that the solution of the *vector* equation for 3-vectors,

$$\mathbf{a}+\mathbf{x} = \mathbf{0}$$

is

$$\mathbf{x} = -\mathbf{a}$$

If two vectors are equal then it means that the vectors are equal in magnitude, are parallel, and have the same sense; but if we are given $\mathbf{a} = -\mathbf{b}$ then the vectors are equal in magnitude, have the same orientation, i.e. parallel but are opposite in sense. (Once

74

more the student is reminded that equality between vector quantities implies that they are of the same nature.)

Addition of three non-coplanar vectors a, b, c.

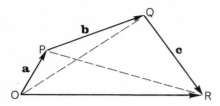

Fig. 76

If **OP** represents vector **a**, then using starting point P **PQ** is drawn equal and parallel to vector **b**, **PQ** represents **b**. Similarly by drawing **QR**, from the terminal point Q of vector **b**, equal and parallel to vector **c** then **QR** represents vector **c**. **OQ** represents **a+b** and **PR** represents **b+c**

$$OR = OQ + c = (a+b) + c$$

But

$$OR = OP + PR = a + (b+c)$$

Hence

$$(a+b) + c = a + (b+c)$$

The multiplication of a vector quantity by a scalar number, treated on pages 16 to 19 is equally valid for 3-vectors as for 2-vectors and the Laws of Vector Algebra stated on page 25 will therefore remain valid in this extension to three dimensions.

The development of vector algebra received its main urge from the demands of physicists and engineers; mathematicians were slow to appreciate the power of this branch of mathematics. During the nineteenth century, the methods developed were mainly those which were of the greatest use to the practical worker in the applications of the subject at a high level. Where these methods are valuable they are being retained but as subsidiary methods in the general development of an algebra of vectors. In the treatment of 3-vectors algebraic methods are so valuable that they are universally adopted. Just as 2-vectors need *two orthogonal*

4

base vectors for an algebraic treatment, so 3-vectors need three orthogonal base vectors. The advantages of using orthogonal bases are so obvious in most of our work that other bases are often neglected.

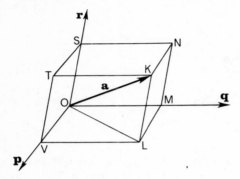

Fig. 77

Consider vector **a** in three dimensions and vectors **p**, **q** and **r** which are non-coplanar in the directions shown. Complete the parallelepiped KLMNSTVO

OK = OL + LK and

OL = OV + VL

OK = OV + VL + LK.

But vector **OV** is parallel to vector **p**

$\Rightarrow$ **OV** = λ**p**

  VL is parallel to vector **q**

  VL = μ**q**

and similarly

LK is parallel to vector **v**

LK = γ**r**

Vector **a** can now be put into its *component* vectors

a = **OK** = λ**p** + μ**q** + γ**r**

where λ, μ and γ are pure numbers, i.e. scalars.

Since only one parallelepiped can be constructed on the diagonal OK with sides parallel to the base vectors, then the representation of vector **a** in terms of **p**, **q** and **r** must be unique.

It follows that any four vectors in a space of three dimensions must be linearly dependent vectors just as it was shown earlier that any three vectors in a plane must also be linearly dependent.

The vector **a** has been expressed in terms of the three base vectors **p**, **q** and **r**. It is obvious that to have three arbitrary vectors in varying directions would present many problems, and for the sake of uniformity a standard group of *base* vectors would need to be defined. Since the three orthogonal right-handed system of axes has already become standard in mathematics, physics, engineering and so on, it is obvious that vector algebra would become more applicable on the same standards.

Now consider the same vector **a** as before but on a right handed orthogonal base.

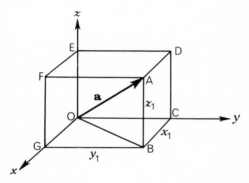

Fig. 78

The parallelepiped has now become a cuboid giving ease of calculation

$$\mathbf{a} = \mathbf{OA} = \mathbf{OB} + \mathbf{BA}$$

$$= (\mathbf{OC} + \mathbf{CB}) + \mathbf{BA}$$

$$= \mathbf{OC} + \mathbf{OG} + \mathbf{OE}$$

$$= \mathbf{OG} + \mathbf{OC} + \mathbf{OE}$$

$$= x_1\hat{\mathbf{x}} + y_1\hat{\mathbf{y}} + z_1\hat{\mathbf{z}}$$

Having adopted the standard right handed system of axes for

general work the unit vectors $\hat{x}$, $\hat{y}$ and $\hat{z}$ are given special symbols, $\hat{x}$ is replaced by $\mathbf{i}$, $\hat{y}$ by $\mathbf{j}$ and $\hat{z}$ by $\mathbf{k}$

So now we can express $\mathbf{a}$ as $x_1\mathbf{i} + y_1\mathbf{j} + z_1\mathbf{k}$

$$\mathbf{a} = x_1\mathbf{i} + y_1\mathbf{j} + z_1\mathbf{k}$$

x_1, y_1, z_1, are the magnitudes of the three components along the three base directions.

The three unit vectors

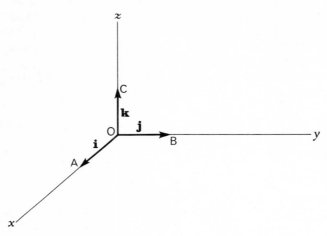

Fig. 79

The three unit vectors $\mathbf{i}$, $\mathbf{j}$ and $\mathbf{k}$ have unit magnitude. In the diagram **OA** represents $\mathbf{i}$ and length of OA = 1 unit. **OB** and **OC**

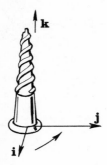

Fig. 80

represent $\mathbf{j}$ and $\mathbf{k}$. Similarly, length of OB = 1 unit and length of OC = 1 unit. A is the point $(1, 0, 0)$, B is $(0, 1, 0)$ and C is $(0, 0, 1)$.

The three unit base orthogonal vectors form a right handed triple in the order $\mathbf{i}, \mathbf{j}, \mathbf{k}$. This is shown by the motion of the wood screw. Any vector along the direction of $\mathbf{i}$ that is along Ox can be represented as $k\mathbf{i}$. If the length of the vector along Ox is x_1, then the vector is $x_1\mathbf{i}$. Similarly vectors along Oy and Oz can be represented as $y_1\mathbf{j}$ and $z_1\mathbf{k}$ if y_1 and z_1 are their magnitudes.

78

Since we are dealing with free vectors the starting point of the vector displacement can be located at any point, not necessarily their origin.

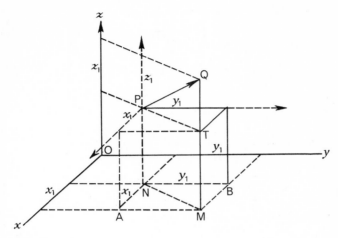

Fig. 81

If **PQ** represents the vector **a**, from P draw three axes parallel to Ox, Oy and Oz. PN is $\perp^r$ to plane xy. QM is also $\perp^r$ to plane xy and PT is $\perp^r$ to QM.

For the axes with origin at P, x_1, y_1 and z_1 are the components of **PQ** on the x, y and z axes originating at P.

For the axes origin at O, x_1, y_1 and z_1, will still be the components of **PQ** in these x, y and z axes, since the vectors are free vectors and a vector equal to **PQ** having its starting point at O would have the same components on the axes through O as **PQ** has in the axes through P. Hence:

$$\mathbf{a} = x_1\mathbf{i} + y_1\mathbf{j} + z_1\mathbf{k}$$

Vector joining two points in space

In Figure 82 the coordinates of S are

$$x_S = OB = 2; \, y_S = OA = 2; \, z_S = SP = 3$$

and the coordinates of T are

$$x_T = OC = 4; \, y_T = OD = 5; \, z_T = TQ = 6$$

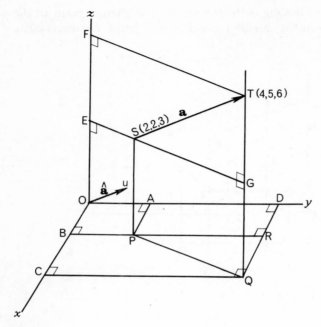

Fig. 82

Component of **ST** in direction $Ox = BC = 4 - 2 = 2$

„ „ **ST** „ „ $Oy = AD = 5 - 2 = 3$

„ „ **ST** „ „ $Oz = EF = 6 - 3 = 3$

In component form

$$\mathbf{ST} = 2\mathbf{i} + 3\mathbf{j} + 3\mathbf{k}$$

Magnitude of **ST**

$$ST^2 = TG^2 + SG^2 \quad \text{using the Theorem of Pythagoras}$$
$$= PQ^2 + TG^2$$
$$= QR^2 + PR^2 + TG^2$$
$$= CB^2 + AD^2 + EF^2$$
$$= 2^2 + 3^2 + 3^2$$
$$= 22$$
$$|\mathbf{ST}| = \sqrt{22}$$

80

Later another method of finding |**ST**| will be developed. It is seen that |**ST**| = $\sqrt{}$ sum of the squares of the components and later this will be demonstrated again using the scalar product.

Putting **ST** = **a** then this can be restated using the unit vector parallel to **ST** as

$$\mathbf{a} = |\mathbf{a}| \cdot \hat{\mathbf{a}} = \sqrt{22} \cdot \hat{\mathbf{a}}$$

giving

$$\hat{\mathbf{a}} = \frac{\mathbf{a}}{\sqrt{22}} = \frac{2\mathbf{i}}{\sqrt{22}} + \frac{3\mathbf{j}}{\sqrt{22}} + \frac{3\mathbf{k}}{\sqrt{22}}$$

showing this unit vector $\hat{\mathbf{a}}$ with its initial point at the origin on the diagram is **OU**, then the cosines of the angles made by **OU** with the base directions are called the Direction Cosines of vector **ST**.

Since the unit vector has unit length the direction cosines are the components of the unit vector in the base directions:

i.e. $\cos \lfloor \text{UOB} = \dfrac{2}{\sqrt{22}}$ or $\dfrac{2\sqrt{22}}{22} = \dfrac{\sqrt{22}}{11}$

$\cos \lfloor \text{UOA} = \dfrac{3}{\sqrt{22}}$ or $\dfrac{3\sqrt{22}}{22}$

$\cos \lfloor \text{UOE} = \dfrac{3}{\sqrt{22}}$ or $\dfrac{3\sqrt{22}}{22}$

Since **ST** is parallel to **OU** then these are also the direction cosines of the vector **ST**.

Representation by components in column matrix form

This means that a 3-vector in space could be specified completely by the three components in a column matrix (or column vector) $\begin{bmatrix} x_1 \\ y_1 \\ z_1 \end{bmatrix}$

The addition of two vectors **a** and **b** could be carried out by the method on page 44 by the addition of the components, i.e. the *elements* of the column matrix.

If $\mathbf{a} = \begin{bmatrix} x_1 \\ y_1 \\ z_1 \end{bmatrix}$ and $\mathbf{b} = \begin{bmatrix} x_2 \\ y_2 \\ z_2 \end{bmatrix}$

Then $\mathbf{a} + \mathbf{b} = \begin{bmatrix} (x_1 + x_2) \\ (y_1 + y_2) \\ (z_1 + z_2) \end{bmatrix}$

This is shown in the diagram below $\mathbf{OP} + \mathbf{PQ} = \mathbf{OQ}$

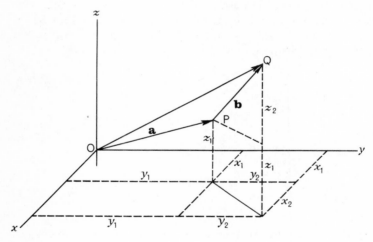

Fig. 83

It is seen that $\mathbf{a} - \mathbf{b} = \begin{bmatrix} (x_1 - x_2) \\ (y_1 - y_2) \\ (z_1 - z_2) \end{bmatrix}$

Since $\mathbf{a} = \begin{bmatrix} x_1 \\ y_1 \\ z_1 \end{bmatrix}$ $-\mathbf{b} = \begin{bmatrix} -x_2 \\ -y_2 \\ -z_2 \end{bmatrix}$

Representation in component vector form

$$\mathbf{a} = x_1\mathbf{i} + y_1\mathbf{j} + z_1\mathbf{k}$$

$$\mathbf{b} = x_2\mathbf{i} + y_2\mathbf{j} + z_2\mathbf{k}$$

$$\mathbf{a} + \mathbf{b} = x_1\mathbf{i} + y_1\mathbf{j} + z_1\mathbf{k} + x_2\mathbf{i} + y_2\mathbf{j} + z_2\mathbf{k}$$

$$= (x_1 + x_2)\mathbf{i} + (y_1 + y_2)\mathbf{j} + (z_1 + z_2)\mathbf{k}$$

and

$$\mathbf{a} - \mathbf{b} = (x_1 - x_2)\mathbf{i} + (y_1 - y_2)\mathbf{j} + (z_1 - z_2)\mathbf{k}$$

Multiplication of a 3-vector by a scalar

If $\mathbf{a} = x_1\mathbf{i} + y_1\mathbf{j} + z_1\mathbf{k}$ and m is a scalar then

$$m\mathbf{a} = mx_1\mathbf{i} + my_1\mathbf{j} + mz_1\mathbf{k}$$

or in matrix form

$$m\mathbf{a} = \begin{bmatrix} mx_1 \\ my_1 \\ mz_1 \end{bmatrix} = m\begin{bmatrix} x_1 \\ y_1 \\ z_1 \end{bmatrix}$$

The laws of vector algebra for addition and scalar multiplication of 2-vectors on page 25 are equally true for 3-vectors, and can be verified in a similar manner.

Scalar or dot products of the base vectors i, j, and k

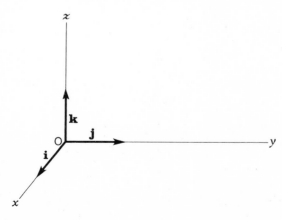

Fig. 84

4*

$\mathbf{i} \cdot \mathbf{i} = |\mathbf{i}||\mathbf{i}| \cos 0° = |\mathbf{i}|^2 = 1$

$\mathbf{j} \cdot \mathbf{j} = |\mathbf{j}||\mathbf{j}| \cos 0° = |\mathbf{j}|^2 = 1$

$\mathbf{k} \cdot \mathbf{k} = |\mathbf{k}||\mathbf{k}| \cos 0° = |\mathbf{k}|^2 = 1$

$\mathbf{i} \cdot \mathbf{j} = |\mathbf{i}| \cdot |\mathbf{j}| \cos 90° = 0$

Similarly $\mathbf{j} \cdot \mathbf{k} = 0$ and $\mathbf{k} \cdot \mathbf{i} = 0$.

These scalar products can be shown in tabular form:

	0	1	i	j	k
0	0	0	0	0	0
1	0	1	i	j	k
i	0	i	1	0	0
j	0	j	0	1	0
k	0	k	0	0	1

Table 1

The table shows how *the product* of any two elements of the set 0, 1, **i**, **j** and **k** always forms a member of the same set. Notice too that the set of elements **i**, **j**, **k** by themselves give a table of products which is the UNIT MATRIX (shown by the thicker lines in the table).

On page 65 it was shown that the *dot* product or scalar product of coplanar vectors is *commutative and distributive*, we now extend this to three dimensions.

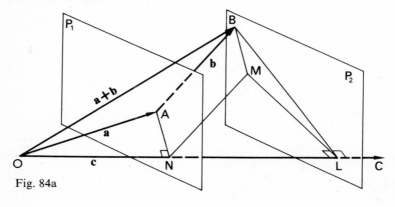

Fig. 84a

84

To show that $\mathbf{c} \cdot (\mathbf{a} + \mathbf{b}) = \mathbf{c} \cdot \mathbf{a} + \mathbf{c} \cdot \mathbf{b}$ in three dimensions: In Fig. 84a OC, OA, AB and OB represents vectors $\mathbf{c}$, $\mathbf{a}$, $\mathbf{b}$ and $\mathbf{a} + \mathbf{b}$. P_1 and P_2 are planes perpendicular to the direction OC, P_1 and P_2 pass through the end-points of vectors OA, AB, A and B respectively.

Since planes P_1 and P_2 are perpendicular to the direction OC then AN is perpendicular to OC and ON is the projection of vector $\mathbf{a}$ on vector $\mathbf{c}$. NM is parallel to AB and therefore the intercepts AB and NM cut off by the parallel planes P_1 and P_2 are equal in length. Hence NM is equal in length to AB and parallel to it, and NM also represents vector $\mathbf{b}$. Since ML is perpendicular to OC, NL is the projection of vector $\mathbf{b}$ on vector $\mathbf{c}$ and since BL is perpendicular to vector $\mathbf{c}$ then OL is the projection of vector $(\mathbf{a} + \mathbf{b})$ on vector $\mathbf{c}$.

Therefore, $\mathbf{c} \cdot (\mathbf{a} + \mathbf{b}) = |\mathbf{c}| \cdot \text{OL}$
$$= |\mathbf{c}| \cdot (\text{ON} + \text{NL}) \text{ since O, N, and L are collinear}$$
$$= |\mathbf{c}| \cdot \text{ON} + |\mathbf{c}| \cdot \text{NL}$$
$$= \mathbf{c} \cdot \mathbf{a} + \mathbf{c} \cdot \mathbf{b}$$

We can now use the DISTRIBUTIVE LAW to find the scalar product of two vectors given in component form.

$$\mathbf{a} \cdot \mathbf{b} = (x_1\mathbf{i} + y_1\mathbf{j} + z_1\mathbf{k}) \cdot (x_2\mathbf{i} + y_2\mathbf{j} + z_2\mathbf{k})$$
$$= x_1\mathbf{i} \cdot (x_2\mathbf{i} + y_2\mathbf{j} + z_2\mathbf{k}) + y_1\mathbf{j} \cdot (x_2\mathbf{i} + y_2\mathbf{j} + z_2\mathbf{k})$$
$$+ z_1\mathbf{k} \cdot (x_2\mathbf{i} + y_2\mathbf{j} + z_2\mathbf{k})$$
$$= x_1x_2 + y_1y_2 + z_1z_2$$

since all other products are zero by the table above. This result could have been obtained by the matrix multiplication since

$$\mathbf{a} = \begin{bmatrix} x_1 \\ y_1 \\ z_1 \end{bmatrix} \quad \mathbf{b} = \begin{bmatrix} x_2 \\ y_2 \\ z_2 \end{bmatrix}$$

Putting $\mathbf{a} = [x_1 \ y_1 \ z_1]$ i.e. in Row vector form gives

$$\mathbf{a} \cdot \mathbf{b} = x_1x_2 + y_1y_2 + z_1z_2$$

This is also called the INNER PRODUCT.

Magnitude of a vector

Now suppose $\mathbf{b} = \mathbf{a}$ so that $x_1 = x_2$, $y_1 = y_2$, $z_1 = z_2$. Then

$$\mathbf{a} \cdot \mathbf{a} = x_1^2 + y_1^2 + z_1^2$$

But

$$\mathbf{a} \cdot \mathbf{a} = |\mathbf{a}||\mathbf{a}|\cos 0° = |\mathbf{a}|^2$$
$$\Rightarrow |\mathbf{a}| = \sqrt{x_1^2 + y_1^2 + z_1^2}$$
$$\Rightarrow |\mathbf{b}| = \sqrt{x_2^2 + y_2^2 + z_2^2}$$

Angle between two vectors a and b

$$\mathbf{a} = x_1\mathbf{i} + y_1\mathbf{j} + z_1\mathbf{k}$$
$$\mathbf{b} = x_2\mathbf{i} + y_2\mathbf{j} + z_2\mathbf{k}$$
$$|\mathbf{a}| = \sqrt{x_1^2 + y_1^2 + z_1^2}$$
$$|\mathbf{b}| = \sqrt{x_2^2 + y_2^2 + z_2^2}$$
$$\mathbf{a} \cdot \mathbf{b} = x_1x_2 + y_1y_2 + z_1z_2$$

also

$$\mathbf{a} \cdot \mathbf{b} = |\mathbf{a}||\mathbf{b}|\cos\theta$$

where θ is the angle between the vectors.

$$\Rightarrow \cos\theta = \frac{x_1x_2 + y_1y_2 + z_1z_2}{|\mathbf{a}||\mathbf{b}|}$$

$$= \frac{x_1x_2 + y_1y_2 + z_1z_2}{\sqrt{x_1^2 + y_1^2 + z_1^2} \cdot \sqrt{x_2^2 + y_2^2 + z_2^2}}$$

Scalar product with a unit vector

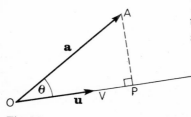

Fig. 85

Let vector $\mathbf{a}$ make an angle θ with the unit vector $\mathbf{u}$

$$\mathbf{a} \cdot \mathbf{u} = |\mathbf{a}||\mathbf{u}|\cos\theta \quad \text{but} \quad |\mathbf{u}| = 1$$

$$= |\mathbf{a}|\cos\theta$$

$$= \text{OP}$$

= projection of **OA** on the direction of **u**

Component form using scalar product

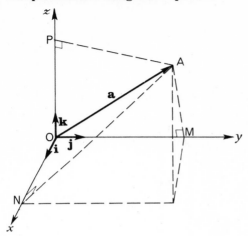

Fig. 86

The projection of vector **a** in the direction of the *x*-axis has been shown to be the scalar product of **a** with a unit vector in the direction of the *x*-axis. Since **i** is the unit vector in the direction O*x* then

$$x = \text{ON} = \mathbf{a} \cdot \mathbf{i}$$

Similarly

$$y = \text{OM} = \mathbf{a} \cdot \mathbf{j} \quad \text{and} \quad z = \text{OP} = \mathbf{a} \cdot \mathbf{k}$$

Now $\mathbf{a} = x\mathbf{i} + y\mathbf{j} + z\mathbf{k}$.
This can now be put in the alternative form

$$\mathbf{a} = (\mathbf{a} \cdot \mathbf{i})\mathbf{i} + (\mathbf{a} \cdot \mathbf{j})\mathbf{j} + (\mathbf{a} \cdot \mathbf{k})\mathbf{k}$$

Using matrix notation the vector **a** can also be expressed as the product of the two matrices

$$\mathbf{a} = \begin{bmatrix} x & y & z \end{bmatrix} \begin{bmatrix} \mathbf{i} \\ \mathbf{j} \\ \mathbf{k} \end{bmatrix} = x\mathbf{i} + y\mathbf{j} + z\mathbf{k}$$

Example 1
Find the projection of vector $\mathbf{a} = 2\mathbf{i} + 3\mathbf{j} + 2\mathbf{k}$ in the direction of vector $\mathbf{b} = 1\mathbf{i} + 2\mathbf{j} + 1\mathbf{k}$.

An Introduction to Vectors

The magnitude of vector $\mathbf{b} = \sqrt{1^2 + 2^2 + 1^2} = \sqrt{6}$. The unit vector in the direction of $\mathbf{b}$ is $\hat{\mathbf{b}}$ and

$$\mathbf{b} = |\mathbf{b}|\hat{\mathbf{b}} = \sqrt{6}\hat{\mathbf{b}}$$
$$\Rightarrow \hat{\mathbf{b}} = \frac{1\mathbf{i}}{\sqrt{6}} + \frac{2\mathbf{j}}{\sqrt{6}} + \frac{1\mathbf{k}}{\sqrt{6}}$$

Now the projection of $\mathbf{a}$ in the direction $\hat{\mathbf{b}}$ is given by $\mathbf{a} \cdot \hat{\mathbf{b}}$ and

$$\mathbf{a} \cdot \hat{\mathbf{b}} = (2\mathbf{i} + 3\mathbf{j} + 2\mathbf{k}) \cdot \left(\frac{1\mathbf{i}}{\sqrt{6}} + \frac{2\mathbf{j}}{\sqrt{6}} + \frac{1\mathbf{k}}{\sqrt{6}} \right)$$
$$= \frac{2}{\sqrt{6}} + \frac{6}{\sqrt{6}} + \frac{2}{\sqrt{6}}$$
$$= \frac{10}{\sqrt{6}} = \frac{10\sqrt{6}}{6} = \frac{5\sqrt{6}}{3}.$$

Hence the projection of vector $\mathbf{a}$ in the direction of vector $\mathbf{b}$ is $\frac{5}{3} \cdot \sqrt{6}$ units long.

Example 2

If vector $\mathbf{p} = 2\hat{\mathbf{x}} + 3\hat{\mathbf{y}}$ and vector $\mathbf{q} = 3\hat{\mathbf{x}} + 2\hat{\mathbf{y}}$ find the projection of $\mathbf{p}$ in the direction of $\mathbf{q}$.

The magnitude of vector $\mathbf{q}$ is $\sqrt{3^2 + 2^2} = \sqrt{13}$. The unit vector $\hat{\mathbf{q}}$ in the direction $\mathbf{q}$

$$= \frac{\mathbf{q}}{|\mathbf{q}|}$$
$$= \frac{\mathbf{q}}{\sqrt{13}}$$

Projection of $\mathbf{p}$ on $\mathbf{q} = \mathbf{p} \cdot \hat{\mathbf{q}}$

$$= (2\hat{\mathbf{x}} + 3\hat{\mathbf{y}}) \cdot \frac{(3\hat{\mathbf{x}} + 2\hat{\mathbf{y}})}{\sqrt{13}}$$
$$= \frac{6+6}{\sqrt{13}} = \frac{12}{\sqrt{13}} = \frac{12\sqrt{13}}{13}$$

In future, when necessary, the length of a projection of a vector in a given direction can always be expressed as the scalar product of the vector with a unit vector on the given direction.

The equation of a line perpendicular to a given vector

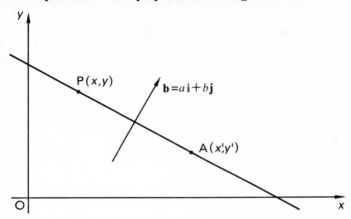

Fig. 86a

Let the given vector be $\mathbf{b} = a\mathbf{i}+b\mathbf{j}$, let the required line perpendicular to vector $\mathbf{b}$ pass through the given point A(x', y'). If P(x, y) is any point on the line as shown, then vector $\mathbf{AP} = (x-x')\mathbf{i}+(y-y')\mathbf{j}$. If the line AP is to be perpendicular to the direction of vector $\mathbf{b}$, then $\mathbf{AP}.\mathbf{b} = 0$ is the required condition, giving

$$[(x-x')\mathbf{i}+(y-y')\mathbf{j}].(a\mathbf{i}+b\mathbf{j}) = 0$$

which reduces to

$$a(x-x')+b(y-y') = 0$$

giving $ax+by = ax'+by'$

but since a, b, x', y' are all constants we can replace $ax'+by'$ by a constant K and the expression now becomes

$$ax+by = K$$

and if $-K$ is replaced by the constant c then

$$ax+by+c = 0.$$

By using a similar method we can show that one of the vectors perpendicular to the line $ax+by+c = 0$ is $\mathbf{p} = a\mathbf{i}+b\mathbf{j}$. So, by inspection, we can state the vector $2\mathbf{i}+3\mathbf{j}$ is perpendicular to the line $2x+3y-4 = 0$. From the properties of the perpendicular vector we can find the perpendicular distance from a given point to a given

89

straight line, and the angle between two lines, using the Scalar product.

The angle between two intersecting straight lines

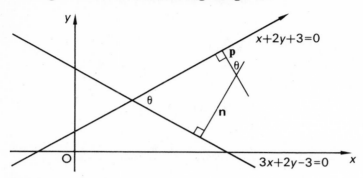

Fig. 86b

From the diagram it is clear that the acute angle θ between the given lines is equal to the acute angle between the directions of the perpendicular vectors. The vector perpendicular to the line $x+2y-3=0$ is, from previous work, $\mathbf{p} = 1\mathbf{i}+2\mathbf{j}$ and likewise $\mathbf{q} = 3\mathbf{i}+2\mathbf{j}$ is the vector perpendicular to the line $3x+2y-3=0$. Now since $\mathbf{p}\cdot\mathbf{q} = |\mathbf{p}||\mathbf{q}| \cos \theta$, then

$$\cos \theta = \frac{\mathbf{p}\cdot\mathbf{q}}{|\mathbf{p}||\mathbf{q}|}$$

$$= \frac{7}{\sqrt{5}\cdot\sqrt{13}}$$

$$= \frac{7}{\sqrt{65}}$$

$$= \frac{7\sqrt{65}}{65}$$

Hence, θ is $29° \ 42'$ and this is the acute angle between the given lines.

To find the perpendicular distance from a given point to a given line using the perpendicular vector

To find the component of a vector in a given direction we need to find the scalar product of the vector with a unit vector in the given direction.

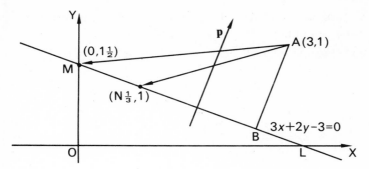

Fig. 86c

It is required to find the distance from the point A(3, 1) along the perpendicular AB, to the line $3x + 2y - 3 = 0$. A vector perpendicular to the line is $\mathbf{p} = 3\mathbf{i} + 2\mathbf{j}$ and BA is parallel to this vector. A unit vector in the direction of $\mathbf{p}$ is

$$\hat{\mathbf{p}} = \frac{\mathbf{p}}{|\mathbf{p}|}.$$

Choosing any convenient point of the line LM, say M(0, $1\frac{1}{2}$), then the scalar product of the vector $\mathbf{AM}$ with the unit vector $\hat{\mathbf{p}}$ gives the component in the perpendicular direction, i.e., the length AB.

Now $\hat{\mathbf{p}} = \dfrac{1}{\sqrt{13}}(3\mathbf{i} + 2\mathbf{j})$ and $\mathbf{AM} = (0-3)\mathbf{i} + (1\frac{1}{2} - 1)\mathbf{j}$

$$= -3\mathbf{i} + \tfrac{1}{2}\mathbf{j}$$

$$\mathbf{AM} . \hat{\mathbf{p}} = (-3\mathbf{i} + \tfrac{1}{2}\mathbf{j}) . \left(\frac{1}{\sqrt{13}}(3\mathbf{i} + 2\mathbf{j}) \right)$$

$$= \frac{1}{\sqrt{13}} \cdot (-9 + 1)$$

$$= \frac{-8}{\sqrt{13}}$$

$$= \frac{-8\sqrt{13}}{13}$$

The negative sign indicates that the component is in the direction from A to B since $\hat{\mathbf{p}}$ is in the direction from B to A. Had the point

N($\frac{1}{3}$, 1) been chosen instead of the more convenient point M, the final result would have been just the same since

$$\mathbf{AN} = (\tfrac{1}{3} - 3)\mathbf{i} + (1 - 1)\mathbf{j}$$
$$= -2\tfrac{2}{3}\mathbf{i} + 0\mathbf{j}$$

an $\mathbf{AN} \cdot \mathbf{\hat{p}} = (-2\tfrac{2}{3}\mathbf{i} + 0\mathbf{j}) \cdot \dfrac{1}{\sqrt{13}} (3\mathbf{i} + 2\mathbf{j}$

$$= \dfrac{-8}{\sqrt{13}}$$

$$= \dfrac{-8\sqrt{13}}{13}$$

Exercise 9.

1. $\mathbf{a} = 2\mathbf{i} + 3\mathbf{j} + 4\mathbf{k}$, $\mathbf{b} = \mathbf{i} - 2\mathbf{j} + 3\mathbf{k}$ and $\mathbf{c} = 3\mathbf{i} + \mathbf{j} - 2\mathbf{k}$. Find $(\mathbf{b} + \mathbf{c})$, then find $\mathbf{a} \cdot \mathbf{b}$, $\mathbf{a} \cdot \mathbf{c}$ and $\mathbf{a} \cdot (\mathbf{b} + \mathbf{c})$. State your conclusion.

2. $\mathbf{a} = 3\mathbf{i} + 4\mathbf{j}$ and $\mathbf{b} = 5\mathbf{i} + 6\mathbf{j}$. Find $\mathbf{a} \cdot \mathbf{b}$ and hence find the angle between $\mathbf{a}$ and $\mathbf{b}$.

3. $\mathbf{p} = 4\mathbf{i} + 5\mathbf{j} + \mathbf{k}$, $\mathbf{q} = 3\mathbf{i} + 2\mathbf{j} + 3\mathbf{k}$ and $\mathbf{r} = 2\mathbf{i} + 2\mathbf{j} + 3\mathbf{k}$. Find $(\mathbf{p} - \mathbf{q})$, $\mathbf{r} \cdot \mathbf{p}$, $\mathbf{r} \cdot \mathbf{q}$, and $\mathbf{r} \cdot (\mathbf{p} - \mathbf{q})$. State your conclusion.

4. Prove that

$$(\mathbf{a} + \mathbf{b}) \cdot (\mathbf{a} - \mathbf{b}) = |\mathbf{a}|^2 - |\mathbf{b}|^2$$

5. With the components given in Qu. 3 find $\mathbf{r} \cdot (\mathbf{p} + \mathbf{q})$ and $(\mathbf{r} + \mathbf{p}) \cdot \mathbf{q}$.

6. Given $\mathbf{a} = 2\mathbf{i} + 1\mathbf{j} + 3\mathbf{k}$; $\mathbf{b} = 2\mathbf{i} - 2\mathbf{j} + 3\mathbf{k}$; $\mathbf{c} = 3\mathbf{i} + 2\mathbf{j} - 4\mathbf{k}$. Find (i) $\mathbf{a} + \mathbf{b}$ (ii) $(\mathbf{a} + \mathbf{b}) + \mathbf{c}$ (iii) $\mathbf{b} + \mathbf{c}$ (iv) $\mathbf{a} + (\mathbf{b} + \mathbf{c})$ (v) $3\mathbf{a} + 2\mathbf{b} + 4\mathbf{c}$.

7. For the given vectors in Qu. 6 find $|\mathbf{a}|$, $|\mathbf{b}|$, $|\mathbf{c}|$, $|\mathbf{a} + \mathbf{b}|$.

8. Find the angle between $\mathbf{a}$ and $\mathbf{b}$, $\mathbf{b}$ and $\mathbf{c}$.

9. If $\mathbf{d} = 6\mathbf{i} + 7\mathbf{j} + 3\mathbf{k}$ find scalars p, q, r such that $\mathbf{d} = p\mathbf{a} + q\mathbf{b} + r\mathbf{c}$.

10. Find a unit vector parallel to the vector $\mathbf{a} + \mathbf{b}$.

11. Express vectors $\mathbf{a}$, $\mathbf{b}$ and $\mathbf{c}$ of Qu. 6 in matrix form and rework Qu. 6, 7, 8 using column matrices.

12. Find the projection of $\mathbf{a} = 4\mathbf{\hat{x}} + 5\mathbf{\hat{y}}$ on $\mathbf{b} = 3\mathbf{\hat{x}} + 2\mathbf{\hat{y}}$.

13. Find the projection of vector $\mathbf{p} = 3\mathbf{i} + 4\mathbf{j} + 5\mathbf{k}$ on the direction of vector $\mathbf{q} = 2\mathbf{i} + 1\mathbf{j} + 2\mathbf{k}$.

14. Find the projection of the vector $3\mathbf{p} + 2\mathbf{q}$ (where $\mathbf{p}$ and $\mathbf{q}$ are the vectors given in Qu. 13) on the direction of vector $\mathbf{r} = 2\mathbf{i} + 3\mathbf{j} + 2\mathbf{k}$.

15. Given that $\mathbf{c} \cdot (\mathbf{a} + \mathbf{b}) = \mathbf{c} \cdot \mathbf{a} + \mathbf{c} \cdot \mathbf{b}$ prove that:

(i) $(\mathbf{a}+\mathbf{b}).\mathbf{c} = \mathbf{a}.\mathbf{c}+\mathbf{b}.\mathbf{c}$

(ii) $(\mathbf{a}+\mathbf{b})(\mathbf{c}+\mathbf{d}) = \mathbf{a}.\mathbf{c}+\mathbf{a}.\mathbf{d}+\mathbf{b}.\mathbf{c}+\mathbf{b}.\mathbf{d}$

16. Find the angles made by $\mathbf{r} = x_1\mathbf{i}+y_1\mathbf{j}+{}_1\mathbf{k}$ with the x, y, and z axes.

17.

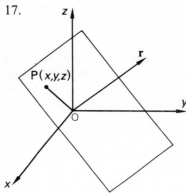

The vector $\mathbf{r} = a\mathbf{i}+b\mathbf{j}+c\mathbf{k}$ is perpendicular to the given plane passing through the origin. Prove that the equation of this plane is $ax+by+cz = 0$. (First take any point P(x, y, z) in the plane and find the scalar product of $\mathbf{r}$ and **OP**.)

Fig. 86d

18. Find the vector $\mathbf{n}$ perpendicular to the vector $\mathbf{p}$ where $\mathbf{p} = a\mathbf{i}+b\mathbf{j}$ and of the same magnitude.

19. Find the vector $\mathbf{m}$ perpendicular to the vector $\mathbf{p}$ where $p = a\mathbf{i}+b\mathbf{j}$, but having three times the magnitude.

20. Find the vector $\mathbf{s}$ lying in the plane $y = 0$ which is perpendicular to the vector $\mathbf{q} = 3\mathbf{i}+4\mathbf{j}+2\sqrt{6}\mathbf{k}$.

Position vectors in parametric form

Consider a point P moving round a circle of radius a units. Taking the centre of the circle as origin O then the position vector **OP** of the point P makes an angle k with the reference line which is the x-axis as shown.

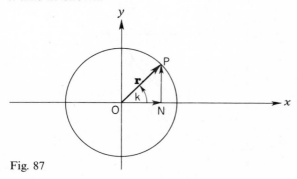

Fig. 87

If **r** represents **OP** then $\mathbf{r} = \mathbf{ON} + \mathbf{NP}$. Since **ON** and **NP** are in the directions of Ox and Oy

$$\mathbf{r} = |\mathbf{r}| \cos k\, \mathbf{i} + |\mathbf{r}| \sin k\, \mathbf{j}$$

but $|\mathbf{r}| = a$

$$\mathbf{r} = (a \cos k)\, \mathbf{i} + (a \sin k)\, \mathbf{j}$$

The position vector **r** is a bound vector with its tail at the fixed point O and is of constant magnitude. The vector **r** has been expressed as a Vector Function of a Real Variable. The tip of vector **r** traces out the circle and if the angle k is a function of time such that $k = \omega t$ then

$$\mathbf{r} = a \cos \omega t\, \mathbf{i} + a \sin \omega t\, \mathbf{j}$$

or $\mathbf{r} = \mathbf{a} \cos \omega t + \mathbf{b} \sin \omega t$

where $\mathbf{a} \cdot \mathbf{b} = 0$

and $|\mathbf{a}| \cdot |\mathbf{b}| = |\mathbf{r}|^2$

and **r** is now given as a vector function of time.

Parametric representation of the ellipse

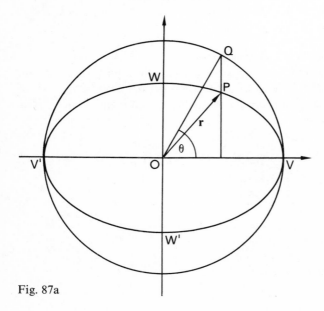

Fig. 87a

For the point P on the ellipse shown in Fig. 87a the position vector **r** is given by

$$\mathbf{r} = \cos\theta\,\mathbf{a} + \sin\theta\,\mathbf{b}$$

where **a** . **b** = 0 and |**a**| > |**b**| and the parameter θ is the eccentric angle in the auxiliary circle, centre O and radius |**a**|.

The centre of the ellipse is situated at the origin O and the major axis $V'V(= 2|\mathbf{a}|)$ is in the direction of the vector **a**. The minor axis $W'W\ (= 2|\mathbf{b}|)$ is perpendicular to $V'V$, since **a** . **b** = 0.

If the origin for the position vectors is also the origin for the co-ordinate system then the form of **r** becomes

$$\mathbf{r} = a\cos\theta\,\mathbf{i} + b\sin\theta\,\mathbf{j}$$

where $a > b$ for the figure shown.

If the point Q moves round the auxiliary circle at constant speed S then $S = a\omega t$, where ω is the angular velocity, and this means that ω is a constant also. The position vector **r** can now be written as a vector function of t

$$\mathbf{r} = a\cos\omega t\,\mathbf{i} + b\sin\omega t\,\mathbf{j}$$

Parametric representation of the parabola

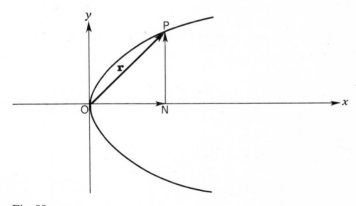

Fig. 88

P is any point on the parabola and taking the vertex of the parabola as origin we have **OP** as the position vector of P. Representing **OP** by **r**, vector addition gives

$$\mathbf{r} = \mathbf{ON} + \mathbf{OP}$$

From the well-known properties of the parabola we can express ON and NP as ap^2 and $2ap$, where the scalar p is a parameter and the scalar constant a determines the parabola, thus

$$\mathbf{r} = ap^2\,\mathbf{i} + 2ap\,\mathbf{j}$$

The position vector $\mathbf{r}$ is expressed as a vector function of the parameter p.

Trajectory of a projectile

The trajectory of a projectile is a parabola and the parameter is the time t.

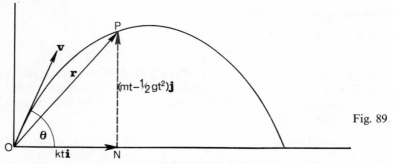

Fig. 89

If the initial velocity of the projectile is $\mathbf{v}$ at angle θ with the horizontal, let the horizontal component of velocity $|\mathbf{v}|\cos\theta = k$ and the vertical component $|\mathbf{v}|\sin\theta = m$. P is the position of the projectile after time t and the position vector $\mathbf{r}$ is given by

$$\mathbf{r} = \mathbf{OP} = \mathbf{ON} + \mathbf{NP}$$
$$= kt\,\mathbf{i} + (mt - \tfrac{1}{2}gt^2)\,\mathbf{j}$$

The circular helix

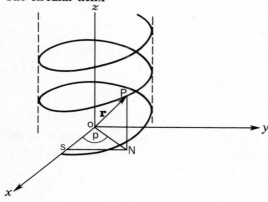

Fig. 90

96

The diagram shows the space curve called the circular helix. The curve grows out of the xy-plane in the shape of a spring touching a cylinder of radius a. The projection on the xy-plane is a circle centre O and if PN is the perpendicular to the xy-plane then

OP = **ON** + **NP**

    = **OS** + **SN** + **NP**

Calling the angle made by ON with Ox p, then NP increases directly with p, i.e.

NP = λp

If **OP** is represented by **r** then

$\mathbf{r} = a \cos p\,\mathbf{i} + a \sin p\,\mathbf{j} + \lambda p\,\mathbf{k}$

This is the vector equation for the circular helix and the position vector **r** is given as a vector function of the scalar p. If the angle p is given as a function of time t, i.e. $p = \omega t$ then the vector function of time is

$\mathbf{r} = a \cos \omega t\,\mathbf{i} + a \sin \omega t\,\mathbf{j} + \lambda \omega t\,\mathbf{k}$

Exercise 10.

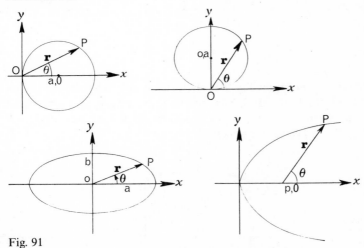

Fig. 91

1. The circle shown has radius a and its centre at $(a, 0)$. Show that OP = $2a \cos \theta$ and express **r** in parametric form.

2. For the circle given of radius a and centre at $(0, a)$, express **r** in parametric form.

3. The ellipse has semi-major axis $= a$ and semi-minor axis $= b$. Express the radius vector **r** in parametric form.

4. For the parabola $y^2 = 4px$, take the focus (p, O) as origin and express the radius vector **r** in parametric form.

5. A plane curve has the equation $\mathbf{r} = c\sqrt{\cot\theta}\,\mathbf{i} + c\sqrt{\tan\theta}\,\mathbf{j}$. Identify the curve and give its equation in cartesian coordinates.

6. A point P moves so that its position vector **r** is given by the vector equation $\mathbf{r} = \mathbf{a} + k\mathbf{p}$ where **a** is a vector of constant magnitude and $d\mathbf{a}/dt = 0$ and k is a scalar variable. What is the locus of P?

7. 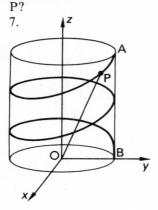 A thin wire is wound into a circular helix which touches a circular cylinder of radius $\sqrt{3}a$ cm and height 4π cm. A bead threaded on the wire is released at A and travels steadily round the wire B at a constant speed of a cm/sec. With the axis shown and the wire making two complete turns find the position vector **r** at time t secs from the commencement of the motion.

Fig. 91a

8. The cylinder of radius $\sqrt{3}a$ is cut by a plane which makes an angle of $30°$ with the horizontal. P is any point on the line of intersection. Give the position vector of P as shown, as a function of θ, and identify the curve round which it moves.

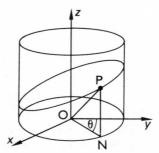

Fig. 91b

9. The position vector of the point P on the curve in the xz plane shown, is given by $\mathbf{r} = a \sec \theta\, \mathbf{i} + (b \tan \theta + h)\mathbf{k}$ where a, b, and h are constants and θ is a variable. Identify the curve and give its equation.

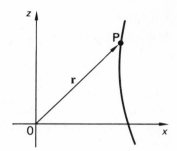

Fig. 91c

10. A particle P is moving so that its position vector $\mathbf{r}$ is given by $\mathbf{r} = k \cos \omega t\, \mathbf{a} + m \sin \omega t\, \mathbf{b}$. Identify the motion of P when
  (i) $\mathbf{a} \cdot \mathbf{b} = 0$
  (ii) $\mathbf{a} \cdot \mathbf{b} = 0$ and $|\mathbf{a}| = |\mathbf{b}|$ and $k = m$
(iii) $\mathbf{a} = \mathbf{b}$

11. A parabola has the equation $y^2 = 4ax$. Give the position vector $\mathbf{r}$ of a point P on the parabola, taking the point $(0, 0)$ as origin.

12. A point P moves so that its position vector $\mathbf{S}$ is given by $\mathbf{S} = m\mathbf{i} + \dfrac{cm}{v}\,\mathbf{j}$ where m and v are variables. What is its motion given that $v = (c\sqrt{m})$ where c is some constant?

Chapter 7
Vector product of two vectors

Vector product

On page 61 it was shown that when two vectors are 'multiplied' together by using algebraic multiplication on the component vectors of each, then the final product is in two distinct parts, the first part is a SCALAR quantity in the plane of the two original vectors and the second part is a vector quantity in a direction at right angles to the plane containing the two vectors. We have dealt with the SCALAR PRODUCT and some of its applications and now we treat the VECTOR PRODUCT, using the sign × to denote the operation.

Definition

If vectors **a** and **b** make an angle of θ with each other, then the VECTOR PRODUCT (or CROSS PRODUCT) is defined as the VECTOR = $|\mathbf{a}||\mathbf{b}| \sin \theta \, \hat{\mathbf{u}}$ where θ is the *smaller* angle measured from **a** to **b**, $\hat{\mathbf{u}}$ is a unit vector at right angles to the plane containing **a** and **b**; **a**, **b** and $\hat{\mathbf{u}}$ form a right-handed system.

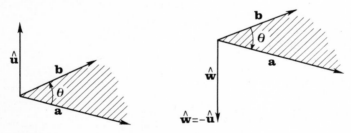

Fig. 92

(i) $\mathbf{a} \times \mathbf{b} = |\mathbf{a}||\mathbf{b}| \sin \theta \hat{\mathbf{u}}$ (ii) $\mathbf{b} \times \mathbf{a} = |\mathbf{b}||\mathbf{a}| \sin \theta \hat{\mathbf{w}}$

But in diagram (ii) the unit vector is opposite in direction to the unit vector of diagram (i) hence

$$\mathbf{a} \times \mathbf{b} = -\mathbf{b} \times \mathbf{a}$$

Vector multiplication is *not* commutative; the *order* of vector multiplication must always be considered.

100

Vector parallelogram

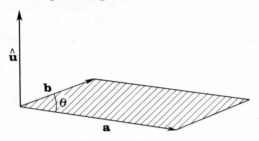

Fig. 93

The vector product of vectors **a** and **b** is

$(|\mathbf{a}||\mathbf{b}|\sin\theta)\hat{\mathbf{u}}$

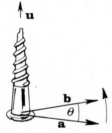

The product $(|\mathbf{a}|\,|\mathbf{b}|\sin\theta)$ is a scalar quantity. It is of course the Area of the *parallelogram* formed by the vectors **a** and **b** at an angle θ.

To emphasise the *order* of multiplication some people speak of the vector product of **a** on **b** to ensure that the angle θ is measured from **a** on to the direction of **b**. We say that the vectors **a**, **b** and $\hat{\mathbf{u}}$ form a right-handed triple. The motion of a right-handed screw illustrates the motion of **a** on **b**.

Fig. 94

Multiplication by a scalar

From the definition of the vector (or cross) product $m(\mathbf{a}\times\mathbf{b}) = m|\mathbf{a}|\,|\mathbf{b}|\sin\theta\hat{\mathbf{u}}$

$|\mathbf{a}|$, $|\mathbf{b}|$ are scalars as is m, so the right hand side of the expression could also be written

$(m|\mathbf{a}|)|\mathbf{b}|\sin\theta\hat{\mathbf{u}}$

or $|\mathbf{a}|(m|\mathbf{b}|)\sin\theta\hat{\mathbf{u}}$

or $|\mathbf{a}|\,|\mathbf{b}|\sin\theta\hat{\mathbf{u}}\,m$

This means that

$m(\mathbf{a}\times\mathbf{b}) = m\mathbf{a}\times\mathbf{b} = \mathbf{a}\times m\mathbf{b} = (\mathbf{a}\times\mathbf{b})m$

101

Magnitude of the vector product
Since $a \times b = |a||b| \sin \theta \hat{u}$ the vector product is a vector of magnitude ($|a||b| \sin \theta$) and this magnitude is also equal to the area of the vector parallelogram. The unit vector $\hat{u}$ gives the direction of the vector product ($a \times b$).

Parallel vectors
1. If a and b are parallel non-zero vectors it follows that $\theta = 0$ when a and b are parallel and $\sin \theta = 0$, hence $a \times b = 0$.
   Here is a condition for parallelism.
2. If we were given initially that $a \times b = 0$ then the conclusion would be either a or b are zero vectors *or* a and b are parallel vectors.

Distributive property of the vector product
If c is a vector perpendicular to both a and b, then c is perpendicular to the plane of a and b by elementary geometry.

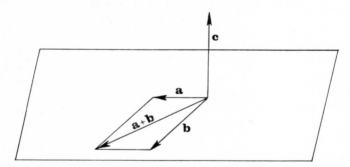

Fig. 95

We shall now show that under this particular condition the vector product

$$c \times (a + b) = c \times a + c \times b$$

Any line at right angles to the direction c lies in the plane of a and b, hence $c \times a$ lies in this plane. In the same way $c \times b$ and $c \times (a + b)$ also lie in this plane.

We show this in the next diagram, c, a and $c \times a$ form a right-handed triple, similarly with c, b and $c \times b$ and with c, $a + b$ and $c \times (a + b)$.

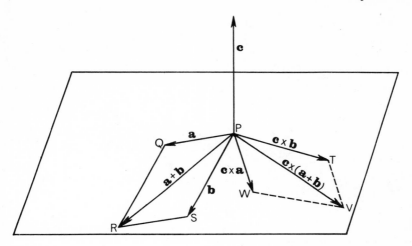

Fig. 96

In the plane perpendicular to $\mathbf{c}$ the vector $\mathbf{c} \times \mathbf{a}$ is at right angles to vector $\mathbf{a}$, the vector $\mathbf{c} \times \mathbf{b}$ is at right angles to vector $\mathbf{b}$ and vector $\mathbf{c} \times (\mathbf{a} + \mathbf{b})$ is at right angles to vector $(\mathbf{a} + \mathbf{b})$.

$$|\mathbf{PT}| = |\mathbf{c}| \, |\mathbf{b}| \sin 90° = |\mathbf{c}| \, |\mathbf{b}| \quad \text{since } \sin 90° = 1$$

$$|\mathbf{PW}| = |\mathbf{c}| \, |\mathbf{a}| \sin 90° = |\mathbf{c}| \, |\mathbf{a}|$$

$$|\mathbf{PV}| = |\mathbf{c}| \, |\mathbf{a} + \mathbf{b}| \sin 90° = |\mathbf{c}| \, |\mathbf{a} + \mathbf{b}|$$

This means that the sides of the figure PTVW are $|\mathbf{c}|$ times the length of the sides of figure PQSR and since the angle TPV = angle RPS, then figure PWVT is a parallelogram with diagonal $\mathbf{PV}$ and $\mathbf{PV} = \mathbf{c} \times (\mathbf{a} + \mathbf{b})$.

But $\quad \mathbf{PV} = \mathbf{PT} + \mathbf{TV} \quad$ and $\quad \mathbf{PT} = (\mathbf{c} \times \mathbf{b}), \quad \mathbf{TV} = \mathbf{PW} = (\mathbf{c} \times \mathbf{a})$
$\Rightarrow \mathbf{c} \times (\mathbf{a} + \mathbf{b}) = \mathbf{c} \times \mathbf{a} + \mathbf{c} \times \mathbf{b}$.
So that under the condition that $\mathbf{c}$ is perpendicular to both $\mathbf{a}$ and $\mathbf{b}$ then the cross product has been shown to be *distributive* with respect to vector multiplication.

Next we consider the general case where $\mathbf{c}$ is not perpendicular to each of the vectors $\mathbf{a}$ and $\mathbf{b}$.

The plane perpendicular to $\mathbf{c}$ is shown with vectors $\mathbf{a}$ and $\mathbf{b}$ inclined at different angles to vector $\mathbf{c}$.

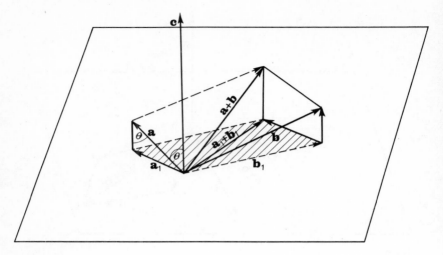

Fig. 97

First we consider the projections of vectors **a**, **b** and (**a** + **b**) in the plane perpendicular to vector **c**. We show these projections **a**₁, **b**₁ and (**a**₁ + **b**₁).

Just as (**a** + **b**) is the sum of **a** and **b** so the sum of the projections **a**₁ and **b**₁ is equal to the projection (**a**₁ + **b**₁).

If θ is the angle between **c** and **a** then $|\mathbf{a}_1| = |\mathbf{a}| \sin \theta$

and $\mathbf{c} \times \mathbf{a} = |\mathbf{c}||\mathbf{a}| \sin \theta \hat{\mathbf{i}}$

but $\mathbf{c} \times \mathbf{a}_1 = |\mathbf{c}||\mathbf{a}_1| \sin 90° \hat{\mathbf{i}} = |\mathbf{c}||\mathbf{a}_1|\hat{\mathbf{i}} = |\mathbf{c}||\mathbf{a}| \sin \theta \hat{\mathbf{i}}$

where $\hat{\mathbf{i}}$ is the unit vector at right angles to **c** and **a** and **a**₁.

This means that $\mathbf{c} \times \mathbf{a} = \mathbf{c} \times \mathbf{a}_1$.

It follows that $\mathbf{c} \times \mathbf{b} = \mathbf{c} \times \mathbf{b}_1$ and $\mathbf{c} \times (\mathbf{a} + \mathbf{b}) = \mathbf{c} \times (\mathbf{a}_1 + \mathbf{b}_1)$. So we can state that

$$\mathbf{c} \times (\mathbf{a} + \mathbf{b}) = \mathbf{c} \times (\mathbf{a}_1 + \mathbf{b}_1)$$

and since vector **c** is perpendicular to (**a**₁ + **b**₁) it was shown that in such a case the distributive law holds;

hence $\mathbf{c} \times (\mathbf{a}_1 + \mathbf{b}_1) = \mathbf{c} \times \mathbf{a}_1 + \mathbf{c} \times \mathbf{b}_1$.

But $\mathbf{c} \times \mathbf{a}_1 = \mathbf{c} \times \mathbf{a}$ and $\mathbf{c} \times \mathbf{b}_1 = \mathbf{c} \times \mathbf{b}$.

104

Hence $c \times (a_1 + b_1) = c \times a + c \times b$.

So finally we state

$$c \times (a+b) = c \times a + c \times b.$$

This is the left-handed distributive law. We have shown that

$$c \times a = -a \times c$$

$$c \times b = -b \times c$$

and $c \times (a+b) = -(a+b) \times c$

so that the right-handed distributive law also holds

$$(a+b) \times c = a \times c + b \times c$$

The distributive laws can be established in different ways but since so much vector analysis which uses component vectors, is dependent on the truth of both the left-handed and right-handed distributive laws, it is essential that their validity is placed beyond dispute.

Vector (or cross) products of the base vectors i, j and k

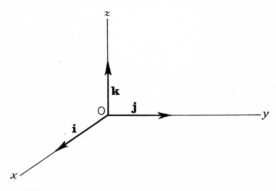

Fig. 98

$i \times i = 0$ $j \times j = 0$ $k \times k = 0$ since in each case the angle between the vectors is $0°$, giving $\sin 0° = 0$. Remembering the adoption of a right-handed system of axes means that although

$$\mathbf{i} \times \mathbf{j} = \mathbf{k}$$

$$\mathbf{j} \times \mathbf{i} = -\mathbf{k}$$

since $\mathbf{i} \times \mathbf{j} = -\mathbf{j} \times \mathbf{i}$

The table shows the vector products of any two vectors of the set **i, j, k**.

	i	j	k
i	0	k	−j
j	−k	0	i
k	j	−i	0

Table 2

Vector product of two vectors in component form

Let $\mathbf{a} = x_1\mathbf{i} + y_1\mathbf{j} + z_1\mathbf{k}$

and $\mathbf{b} = x_2\mathbf{i} + y_2\mathbf{j} + z_2\mathbf{k}$

Then $\mathbf{a} \times \mathbf{b} = (x_1\mathbf{i} + y_1\mathbf{j} + z_1\mathbf{k}) \times (x_2\mathbf{i} + y_2\mathbf{j} + z_2\mathbf{k})$

and we have proved that the vector product is distributive with respect to multiplication, the right hand side will therefore produce nine separate vector products

$$\mathbf{a} \times \mathbf{b} = x_1x_2\mathbf{i} \times \mathbf{i} + x_1\mathbf{i} \times \mathbf{j} + x_1z_2\mathbf{i} \times \mathbf{k} + y_1x_2\mathbf{j} \times \mathbf{i} + y_1y_2\mathbf{j} \times \mathbf{j}$$

$$+ y_1z_2\mathbf{j} \times \mathbf{k} + z_1x_2\mathbf{k} \times \mathbf{i} + z_1y_2\mathbf{k} \times \mathbf{j} + z_1z_2\mathbf{k} \times \mathbf{k}$$

Reference to the table shows that three of these products $\mathbf{i} \times \mathbf{i}$, $\mathbf{j} \times \mathbf{j}$, $\mathbf{k} \times \mathbf{k}$ are all zero and three more are negative ($\mathbf{j} \times \mathbf{i}$, $\mathbf{i} \times \mathbf{k}$, $\mathbf{k} \times \mathbf{j}$).

$$\Rightarrow \mathbf{a} \times \mathbf{b} = x_1y_2\mathbf{k} - x_2y_1\mathbf{k} + x_2z_1\mathbf{j} - x_1z_2\mathbf{j} + y_1z_2\mathbf{i} - y_2z_1\mathbf{i}$$

$$\Rightarrow \mathbf{a} \times \mathbf{b} = (y_1z_2 - y_2z_1)\mathbf{i} + (x_2z_1 - x_1z_2)\mathbf{j} + (x_1y_2 - x_2y_1)\mathbf{k}$$

The right-hand side gives the vector product in component form which makes a right-handed triple with **a** and **b**. This can be illustrated with the motion of a woodscrew turning from **a** to **b**:

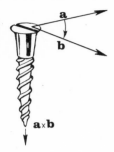

Fig. 99

Vector product in determinant form
The result of the last operation can be rearranged in a form which is then recognisable as a $(3 \times 3)^*$ determinant which provides a very suitable form for memorising.

$$\mathbf{a} \times \mathbf{b} = (y_1 z_2 - y_2 z_1)\mathbf{i} + (x_2 z_1 - x_1 z_2)\mathbf{j} + (x_1 y_2 - x_2 y_1)\mathbf{k}$$

$$= (y_1 z_2 - y_2 z_1)\mathbf{i} - (x_1 z_2 - x_2 z_1)\mathbf{j} + (x_1 y_2 - x_2 y_1)\mathbf{k}$$

$$= \begin{vmatrix} \mathbf{i} & \mathbf{j} & \mathbf{k} \\ x_1 & y_1 & z_1 \\ x_2 & y_2 & z_2 \end{vmatrix}$$

The Rule of Sarrus* if known is also useful for the expansion.
Since the interchange of two rows alters the sign of the determinant and

$$\mathbf{b} \times \mathbf{a} = -\mathbf{a} \times \mathbf{b}$$

$$\mathbf{b} \times \mathbf{a} = \begin{vmatrix} \mathbf{i} & \mathbf{j} & \mathbf{k} \\ x_2 & y_2 & z_2 \\ x_1 & y_1 & z_1 \end{vmatrix}$$

Example
$$\mathbf{a} = 3\mathbf{i} + 2\mathbf{j} + 2\mathbf{k}$$

$$\mathbf{b} = 2\mathbf{i} + 4\mathbf{j} + 3\mathbf{k}$$

* (See expansion of (3×3) determinant in A. E. Coulson *An Introduction to Matrices*, Longmans, 1965.)

5

Find the vector product $\mathbf{a} \times \mathbf{b}$ and $\mathbf{b} \times \mathbf{a}$

(*i*) *Using the determinant method*

$$\mathbf{a} \times \mathbf{b} = \begin{vmatrix} \mathbf{i} & \mathbf{j} & \mathbf{k} \\ 3 & 2 & 2 \\ 2 & 4 & 3 \end{vmatrix}$$

$$= (2.3-4.2)\mathbf{i} - (3.3-2.2)\mathbf{j} + (3.4-2.2)\mathbf{k}$$

$$= -2\mathbf{i} - 5\mathbf{j} + 8\mathbf{k}$$

$$\mathbf{b} \times \mathbf{a} = \begin{vmatrix} \mathbf{i} & \mathbf{j} & \mathbf{k} \\ 2 & 4 & 3 \\ 3 & 2 & 2 \end{vmatrix} = 2\mathbf{i} + 5\mathbf{j} - 8\mathbf{k}$$

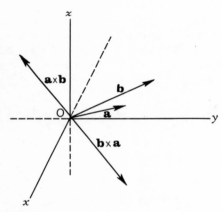

Fig. 100

(*ii*) *Expansion method*

$$\mathbf{a} \times \mathbf{b} = (3\mathbf{i} + 2\mathbf{j} + 2\mathbf{k}) \times (2\mathbf{i} + 4\mathbf{j} + 3\mathbf{k})$$

$$= 3\mathbf{i} \times (2\mathbf{i} + 4\mathbf{j} + 3\mathbf{k}) + 2\mathbf{j} \times (2\mathbf{i} + 4\mathbf{j} + 3\mathbf{k}) + 2\mathbf{k} \times (2\mathbf{i} + 4\mathbf{j} + 3\mathbf{k})$$

$$= 6\mathbf{i} \times \mathbf{i} + 12\mathbf{i} \times \mathbf{j} + 9\mathbf{i} \times \mathbf{k} + 4\mathbf{j} \times \mathbf{i} + 8\mathbf{j} \times \mathbf{j} + 6\mathbf{j} \times \mathbf{k} + 4\mathbf{k} \times \mathbf{i}$$

$$+ 8\mathbf{k} \times \mathbf{j} + 6\mathbf{k} \times \mathbf{k}$$

Reference to Table 2 gives

$$= 12\mathbf{k} - 9\mathbf{j} - 4\mathbf{k} + 6\mathbf{i} + 4\mathbf{j} - 8\mathbf{i}$$

$$= -2\mathbf{i} - 5\mathbf{j} + 8\mathbf{k}$$

$$\mathbf{b} \times \mathbf{a} = (2\mathbf{i} + 4\mathbf{j} + 3\mathbf{k}) \times (3\mathbf{i} + 2\mathbf{j} + 2\mathbf{k})$$

$$= 2\mathbf{i} \times (3\mathbf{i} + 2\mathbf{j} + 2\mathbf{k}) + 4\mathbf{j} \times (3\mathbf{i} + 2\mathbf{j} + 2\mathbf{k}) + 3\mathbf{k} \times (3\mathbf{i} + 2\mathbf{j} + 2\mathbf{k})$$

$$= 6\mathbf{i} \times \mathbf{i} + 4\mathbf{i} \times \mathbf{j} + 4\mathbf{i} \times \mathbf{k} + 12\mathbf{j} \times \mathbf{i} + 8\mathbf{j} \times \mathbf{j} + 8\mathbf{j} \times \mathbf{k} + 6\mathbf{k} \times \mathbf{j} + 6\mathbf{k} \times \mathbf{k}$$

$$= 4\mathbf{k} - 4\mathbf{j} - 12\mathbf{k} + 8\mathbf{i} + 9\mathbf{j} - 6\mathbf{i}$$

$$= 2\mathbf{i} + 5\mathbf{j} - 8\mathbf{k}$$

Example 1
To find $|\mathbf{p} \times \mathbf{q}|$ given $\mathbf{p} = \mathbf{i} - 2\mathbf{j} + 2\mathbf{k}$, $\mathbf{q} = 3\mathbf{i} + \mathbf{j} - 2\mathbf{k}$.
Method 1
$\mathbf{p} \times \mathbf{q} = (\mathbf{i} - 2\mathbf{j} + 2\mathbf{k}) \times (3\mathbf{i} + \mathbf{j} - 2\mathbf{k})$ and since the distributive law has been established,

$$= \mathbf{i} \times (3\mathbf{i} + \mathbf{j} - 2\mathbf{k}) - 2\mathbf{j} \times (3\mathbf{i} + \mathbf{j} - 2\mathbf{k}) + 2\mathbf{k} \times (3\mathbf{i} + \mathbf{j} - 2\mathbf{k})$$

$$= (\mathbf{i} \times \mathbf{j}) - 2(\mathbf{i} \times \mathbf{k}) - 6(\mathbf{j} \times \mathbf{i}) + 4(\mathbf{j} \times \mathbf{k}) + 6(\mathbf{k} \times \mathbf{i}) + 2(\mathbf{k} \times \mathbf{j})$$

Since $\mathbf{i} \times \mathbf{i} = \mathbf{j} \times \mathbf{j} = \mathbf{k} \times \mathbf{k} = 0$, and by reference to Table 2 on page 97, or directly we can now write

$$\mathbf{p} \times \mathbf{q} = \mathbf{k} + 2\mathbf{j} + 6\mathbf{k} + 4\mathbf{i} + 6\mathbf{j} - 2\mathbf{i}$$

$$= 2\mathbf{i} + 8\mathbf{j} + 7\mathbf{k}$$

Hence $|\mathbf{p} \times \mathbf{q}| = \sqrt{2^2 + 8^2 + 7^2} = \sqrt{117}$.
Method 2 Using the determinant method

$$\mathbf{p} \times \mathbf{q} = \begin{vmatrix} \mathbf{i} & \mathbf{j} & \mathbf{k} \\ 1 & -2 & 2 \\ 3 & 1 & -2 \end{vmatrix} = \mathbf{i} \begin{vmatrix} -2 & 2 \\ 1 & -2 \end{vmatrix} - \mathbf{j} \begin{vmatrix} 1 & 2 \\ 3 & -2 \end{vmatrix} + \mathbf{k} \begin{vmatrix} 1 & -2 \\ 3 & 1 \end{vmatrix}$$

$$= \mathbf{i}(+4 - 2) - \mathbf{j}(-2 - 6) + \mathbf{k}(1 + 6)$$

$$= 2\mathbf{i} + 8\mathbf{j} + 7\mathbf{k}$$

then as in Method 1, $|\mathbf{p} \times \mathbf{q}| = \sqrt{117}$.
Example 2
Given $a = 2\mathbf{i} + 3\mathbf{j} + 4\mathbf{k}$ and $\mathbf{b} = 3\mathbf{i} - 2\mathbf{j} + 3\mathbf{k}$ find the angle between vectors $\mathbf{a}$ and $\mathbf{b}$ using the vector product.

$$|\mathbf{a}| = \sqrt{2^2 + 3^2 + 16} = \sqrt{29}$$

$$|\mathbf{b}| = \sqrt{3^2 + 2^2 + 3^2} = \sqrt{22}$$

$$\mathbf{a} \times \mathbf{b} = |\mathbf{a}||\mathbf{b}| \sin\theta\hat{\mathbf{u}} = \sqrt{29}\sqrt{22} \sin\theta\hat{\mathbf{u}}$$

109

But we can also find $\mathbf{a} \times \mathbf{b}$ using the determinant method:

$$\mathbf{a} \times \mathbf{b} = \begin{vmatrix} \mathbf{i} & \mathbf{j} & \mathbf{k} \\ 2 & 3 & 4 \\ 3 & -2 & 3 \end{vmatrix} = 17\mathbf{i} + 6\mathbf{j} - 13\mathbf{k} = |\mathbf{a} \times \mathbf{b}|\hat{\mathbf{u}}$$

But $|\mathbf{a} \times \mathbf{b}| = \sqrt{17^2 + 6^2 + 13^2}$

$$= \sqrt{289 + 36 + 169}$$

$$= \sqrt{494}.$$

Thus $\mathbf{a} \times \mathbf{b} = \sqrt{494}\hat{\mathbf{u}}$

$$\Rightarrow |\mathbf{a}|\,|\mathbf{b}| \sin \theta \hat{\mathbf{u}} = \sqrt{494}\hat{\mathbf{u}}$$

$$\sin \theta = \frac{\sqrt{494}}{\sqrt{29}\sqrt{22}} = \frac{22 \cdot 23}{5 \cdot 385 \times 4 \cdot 69}$$

$$\theta = 1 \cdot 0763 \text{ rad or } 61° \, 40'$$

(The angle between $\mathbf{a}$ and $\mathbf{b}$ could also be determined by using the scalar product.)

Exercise 11.
1. Two coplanar vectors $\mathbf{a}$ and $\mathbf{b}$ make an angle of 30° with each other and $|\mathbf{a}| = |3|$ units, $|\mathbf{b}| = 4$ units. What is the magnitude of their vector product and what is its direction?
2. If $\mathbf{r} = x_1\hat{\mathbf{x}} + y_1\hat{\mathbf{y}}$ and $\mathbf{s} = x_2\hat{\mathbf{x}} + y_2\hat{\mathbf{y}}$, find $\mathbf{r} \times \mathbf{s}$.
3. If $\mathbf{a} = 3\mathbf{i} + 4\mathbf{j}$ and $\mathbf{b} = 2\mathbf{i} + 4\mathbf{j}$, find $\mathbf{a} \times \mathbf{b}$ and hence find the angle between $\mathbf{a}$ and $\mathbf{b}$.
4. The triangle ABC has its vertices determined by the position vectors $\mathbf{a}$, $\mathbf{b}$ and $\mathbf{c}$. Show the area of triangle ABC is

$$\tfrac{1}{2}\{|\mathbf{c} \times \mathbf{b}| + |\mathbf{b} \times \mathbf{a}| + |\mathbf{a} \times \mathbf{c}|\}$$

5. Prove that $(\mathbf{p} + \mathbf{q}) \times (\mathbf{p} - \mathbf{q}) = 2\mathbf{q} \times \mathbf{p}$.
6. Given that $\mathbf{a} \times \mathbf{b} = 0$ and neither $\mathbf{a}$ nor $\mathbf{b}$ equals 0, state your conclusion.
7. If $\mathbf{a} \times \mathbf{b} = 0$ and $\mathbf{a} = 2\mathbf{i} + 3\mathbf{j} + 5\mathbf{k}$, $\mathbf{b} = m\mathbf{i} + n\mathbf{j} + 12\mathbf{k}$ find the values of the scalars m and n.

8. $\mathbf{r} = 3\mathbf{i} + 2\mathbf{j} + 5\mathbf{k}$ and $\mathbf{s} = 2\mathbf{i} + 2\mathbf{j} + 3\mathbf{k}$. Find $\mathbf{r} \cdot \mathbf{s}$ and $\mathbf{r} \times \mathbf{s}$. What is the essential difference between these products?

9. A vector $\mathbf{m}$ has magnitude 5 units and makes an angle of $60°$ with a vector $\mathbf{n}$ of magnitude 4 units. Find the magnitude of their vector product and give its direction.

10. $\mathbf{m} = 2\mathbf{i} + 2\mathbf{j} + 1\mathbf{k}$ and $\mathbf{n} = 4\mathbf{i} + 4\mathbf{j} - 7\mathbf{k}$. Find a unit vector perpendicular to both $\mathbf{m}$ and $\mathbf{n}$.

11. If $\mathbf{a}$ is any vector prove that $\mathbf{a} \times \mathbf{a} = \mathbf{0}$.

12. If $\mathbf{a} = a_1\mathbf{i} + a_2\mathbf{j} + a_3\mathbf{k}$ and $\mathbf{b} = b_1\mathbf{i} + b_2\mathbf{j} + b_3\mathbf{k}$, show that $(\mathbf{a} \times \mathbf{b}) = (a_2b_3 - a_3b_2)\mathbf{i} + (a_3b_1 - a_1b_3)\mathbf{j} + (a_1b_2 - a_2b_1)\mathbf{k}$.

13. If $\mathbf{a} = x_1\mathbf{i} + y_1\mathbf{j} + z_1\mathbf{k}$ and $\mathbf{b} = x_2\mathbf{i} + y_2\mathbf{j} + z_2\mathbf{k}$, show that $|\mathbf{a} \times \mathbf{b}|_2 = (y_1z_2 - z_1y_2)^2 + (z_1x_2 - x_1z_2)^2 + (x_1y_2 - y_1x_2)^2$.

14. Show that $(\mathbf{a} \times \mathbf{b}) \cdot \mathbf{a} = 0$ and $(\mathbf{a} \times \mathbf{b}) \cdot \mathbf{b} = 0$ and state your conclusion.

Chapter 8
Further geometrical applications of vectors

The vectorial approach
The last century has seen a steady change in the approach to geometry: the subject has been made more algebraic and the application of vector methods has been part of the process, which is continuing and gaining momentum. Some reformers would like to see all geometry approached from a vectorial standpoint, others would welcome an even wider algebraic approach; but whatever the viewpoint, vector algebra does provide a better understanding of some branches of the subject, and in analytical geometry of three dimensions it offers powerful alternative methods to the more cumbersome coordinate methods particularly where the scalar or dot product and vector (or cross) product of two vectors can be used to advantage. The use of the scalar (or dot) product in some plane geometry was illustrated on pages 67–69.

First we give some applications of vector (or cross) products before proceeding to three dimensional problems.

The vector parallelogram

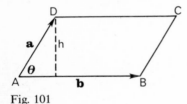

Fig. 101

The parallelogram has vectors **a** and **b** for its sides AD and AB.

The area of ABCD = Base × Altitude

$$= AB \cdot h$$

$$= AB \cdot AD \sin \theta$$

$$= |\mathbf{a}| \, |\mathbf{b}| \sin \theta$$

$$= |(\mathbf{a} \times \mathbf{b})|$$

Area of ABD = $\frac{1}{2}$ parallelogram ABCD = $\frac{1}{2}|\mathbf{a} \times \mathbf{b}|$.

Consider three collinear points A, B and C.

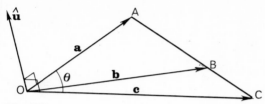

Fig. 102

Take any convenient origin O, then **OA**, **OB** and **OC** are the position vectors of A, B and C. Let these vectors be **a**, **b** and **c** as shown

$$\mathbf{c} \times \mathbf{a} = |\mathbf{c}||\mathbf{a}| \sin \theta \hat{\mathbf{u}}$$

$$= 2(\text{Area of } \triangle OAC)\hat{\mathbf{u}}$$

where $\hat{\mathbf{u}}$ is at right angles to the plane of OAC forming a right handed triple with **OC**, **OA** *in that order.*

Then $\mathbf{a} \times \mathbf{b} = |\mathbf{a}||\mathbf{b}| \sin |AOB| (-\hat{\mathbf{u}})$ since **a, b** and $-\hat{\mathbf{u}}$ form a

$= 2 (\text{Area of } \overline{AOB}) (-\hat{\mathbf{u}})$ right-handed triple also.

and $\mathbf{b} \times \mathbf{c} = 2 (\text{Area of } BOC) (-\hat{\mathbf{u}})$

Hence $(\mathbf{c} \times \mathbf{a}) + (\mathbf{a} \times \mathbf{b}) + (\mathbf{b} \times \mathbf{c}) = 2\hat{\mathbf{u}} (\text{Area } OAC - \text{Area } AOB -$

$\text{Area } BOC)$

$$= \mathbf{0}$$

Hence the condition for three points A, B and C to be collinear is

$$(\mathbf{c} \times \mathbf{a}) + (\mathbf{a} \times \mathbf{b}) + (\mathbf{b} \times \mathbf{c}) = \mathbf{0}$$

Sine formula for $\triangle ABC$

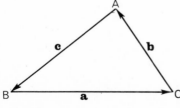

Fig. 103

In the triangle ABC if $\mathbf{a}, \mathbf{b}$ and $\mathbf{c}$ are the sides then

$\mathbf{a} + \mathbf{b} + \mathbf{c} = \mathbf{0}$

the distributive law gives

$\mathbf{a} \times (\mathbf{a} + \mathbf{b} + \mathbf{c}) = \mathbf{0}$

$\Rightarrow (\mathbf{a} \times \mathbf{a}) + (\mathbf{a} \times \mathbf{b}) + (\mathbf{a} \times \mathbf{c}) = \mathbf{0}$

since $\mathbf{a} \times \mathbf{a} = \mathbf{0}$

$$\mathbf{a} \times \mathbf{b} = -\mathbf{a} \times \mathbf{c} = \mathbf{c} \times \mathbf{a} \tag{1}$$

Again

$\mathbf{b} \times (\mathbf{a} + \mathbf{b} + \mathbf{c}) = \mathbf{0}$

$\Rightarrow (\mathbf{b} \times \mathbf{a}) + (\mathbf{b} \times \mathbf{b}) + (\mathbf{b} \times \mathbf{c}) = \mathbf{0}$

since $\mathbf{b} \times \mathbf{b} = \mathbf{0}$

$-\mathbf{b} \times \mathbf{a} = \mathbf{b} \times \mathbf{c}$

or

$$\mathbf{a} \times \mathbf{b} = \mathbf{b} \times \mathbf{c} \tag{2}$$

Combining (1) and (2) gives

$\mathbf{a} \times \mathbf{b} = \mathbf{b} \times \mathbf{c} = \mathbf{c} \times \mathbf{a}$

$(|\mathbf{a}| \, |\mathbf{b}| \sin C)\hat{\mathbf{u}} = (|\mathbf{b}| \, |\mathbf{c}| \sin A)\hat{\mathbf{u}} = (|\mathbf{c}| \, |\mathbf{a}| \sin B)\hat{\mathbf{u}}$ giving

$ab \sin C = bc \sin A = ac \sin B$

Dividing throughout by abc

$$\frac{\sin C}{c} = \frac{\sin A}{a} = \frac{\sin B}{b}$$

which is the SINE RULE for a triangle.

Sum and difference formulae

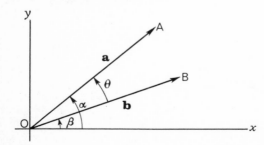

Fig. 104

114

Suppose **a** and **b** are *unit vectors* making angles of α and β with Ox and lying in the plane of Ox and Oy

$$\mathbf{a} = \cos \alpha \mathbf{i} + \sin \alpha \mathbf{j}$$
$$\mathbf{b} = \cos \beta \mathbf{i} + \sin \beta \mathbf{j}$$

The scalar product $\mathbf{a} \cdot \mathbf{b} = |\mathbf{a}||\mathbf{b}| \cos \theta$
$$= 1 \cdot 1 \cdot \cos(\alpha - \beta) \text{ [since } |\mathbf{a}| = |\mathbf{b}| = 1].$$

But $\mathbf{a} \cdot \mathbf{b} = (\cos \alpha \mathbf{i} + \sin \alpha \mathbf{j}) \cdot (\cos \beta \mathbf{i} + \sin \beta \mathbf{j})$
$$= \cos \alpha \cos \beta + \sin \alpha \sin \beta$$
$$\Rightarrow \cos(\alpha - \beta) = \cos \alpha \cos \beta + \sin \alpha \sin \beta$$

Putting $\beta = (-\beta)$
$$\cos(\alpha + \beta) = \cos \alpha \cos \beta - \sin \alpha \sin \beta$$

The vector product of **b** and **a** is $|\mathbf{b}||\mathbf{a}| \sin \theta \mathbf{k} = \sin(\alpha - \beta)\mathbf{k}$ (where **k** is the unit vector forming a right-handed triple with **a** and **b**)

since $\mathbf{b} \times \mathbf{a} = (\cos \beta \mathbf{i} + \sin \beta \mathbf{j}) \times (\cos \alpha \mathbf{i} + \sin \alpha \mathbf{j})$
$$= \cos \beta \sin \alpha \mathbf{i} \times \mathbf{j} + \sin \beta \cos \alpha \mathbf{j} \times \mathbf{i} \qquad \text{since } \mathbf{i} \times \mathbf{j} = \mathbf{k}$$
$$= \sin \alpha \cos \beta \mathbf{k} - \sin \beta \cos \alpha \mathbf{k} \qquad \text{and} \quad \mathbf{j} \times \mathbf{i} = -\mathbf{k}$$
$$= (\sin \alpha \cos \beta - \cos \alpha \sin \beta)\mathbf{k}$$
$$\Rightarrow \sin(\alpha - \beta) = \sin \alpha \cos \beta - \cos \alpha \sin \beta$$

The straight line and the plane in three dimensions
(i) *Equation of a straight line in the analytical form*
Two points determine the position of a straight line. Only one straight line will pass through any two given points.

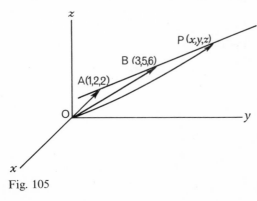

Fig. 105

A is the point $(1, 2, 2)$

B „ „   „ $(3, 5, 6)$

P is any point (x, y, z) on the straight line through A and B. Regarding the line segment **AB** as a vector it has components of 2 in the x direction, 3 in the y direction and 4 in the z direction

$$\mathbf{AB} = 2\mathbf{i} + 3\mathbf{j} + 4\mathbf{k}$$

$$\mathbf{AP} = (x - 1)\mathbf{i} + (y - 2)\mathbf{j} + (z - 2)\mathbf{k}$$

since

$$\mathbf{AP} = m\mathbf{AB}$$

$$(x - 1)\mathbf{i} + (y - 2)\mathbf{j} + (z - 2)\mathbf{k} = m(2\mathbf{i} + 3\mathbf{j} + 4\mathbf{k})$$
$$= 2m\mathbf{i} + 3m\mathbf{j} + 4m\mathbf{k}$$

Equating coefficients gives

$$(x - 1) = 2m$$

$$(y - 2) = 3m$$

$$(z - 2) = 4m$$

$$\Rightarrow \frac{x - 1}{2} = \frac{y - 2}{3} = \frac{z - 2}{4} (= m)$$

This is the required relation between x, y and z

(ii) This relation could be established by another method by using position vectors. Choosing O as the origin for the position vectors

$$\mathbf{AB} = \mathbf{OB} - \mathbf{OA}$$

$$\mathbf{AP} = \mathbf{OP} - \mathbf{OA}$$

$$\mathbf{AB} = \begin{bmatrix} 3 \\ 5 \\ 6 \end{bmatrix} - \begin{bmatrix} 1 \\ 2 \\ 2 \end{bmatrix} = \begin{bmatrix} 2 \\ 3 \\ 4 \end{bmatrix}$$

$$\mathbf{AP} = \begin{bmatrix} x \\ y \\ z \end{bmatrix} - \begin{bmatrix} 1 \\ 2 \\ 2 \end{bmatrix} = \begin{bmatrix} (x - 1) \\ (y - 2) \\ (z - 2) \end{bmatrix}$$

As before $\mathbf{AP} = m\mathbf{AB}$

$$\begin{bmatrix} x - 1 \\ y - 2 \\ z - 2 \end{bmatrix} = m \begin{bmatrix} 2 \\ 3 \\ 4 \end{bmatrix} \text{ giving } \frac{x - 1}{2} = \frac{y - 2}{3} = \frac{z - 2}{4} (= m)$$

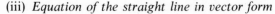

(iii) *Equation of the straight line in vector form*

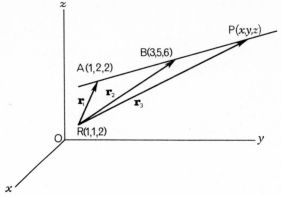

Fig. 106

Any convenient origin can be chosen, let it be the point R(1, 1, 2). The position vectors of A, B and P are $\mathbf{r}_1$, $\mathbf{r}_2$ and $\mathbf{r}_3$.
Then $\mathbf{AB} = \mathbf{r}_2 - \mathbf{r}_1$ and $\mathbf{AP} = \mathbf{r}_3 - \mathbf{r}_1$
Since $\mathbf{AP}$ is parallel to $\mathbf{AB}$
then $\mathbf{AP} = t \cdot \mathbf{AB}$ where t is some variable.

As t varies so the position of P varies *but* always along AB

$$\mathbf{r}_3 - \mathbf{r}_1 = t(\mathbf{r}_2 - \mathbf{r}_1) \tag{i}$$

$$\mathbf{r}_3 - (1-t)\mathbf{r}_1 - t\mathbf{r}_2 = 0$$

This is the equation of the straight line in vector form.

We can reduce this to the analytical form thus

$$\mathbf{r}_3 = (x-1)\mathbf{i} + (y-1)\mathbf{j} + (z-2)\mathbf{k}$$

$$\mathbf{r}_2 = 2\mathbf{i} + 4\mathbf{j} + 4\mathbf{k}$$

$$\mathbf{r}_1 = 0\mathbf{i} + 1\mathbf{j} + 0\mathbf{k}$$

Vector equation (i) can be rewritten

$$(x-1)\mathbf{i} + (y-1)\mathbf{j} + (z-2)\mathbf{k} - 0\mathbf{i} - 1\mathbf{j} - 0\mathbf{k}$$

$$= t(2\mathbf{i} + 4\mathbf{j} + 4\mathbf{k} - 0\mathbf{i} - 1\mathbf{j} - 0\mathbf{k})$$

$$\Rightarrow (x-1)\mathbf{i} + (y-2)\mathbf{j} + (z-2)\mathbf{k} = 2t\mathbf{i} + 3t\mathbf{j} + 4t\mathbf{k}$$

$$\Rightarrow \frac{x-1}{2} = \frac{y-2}{3} = \frac{z-2}{4}$$

117

This shows that the vector equation $r_3 - (1 - t)r_1 - tr_2 = 0$ is equivalent to the analytical form.

Exercise 11a

1. A straight line through the point (p, q, r) is parallel to the vector $ai + bj + ck$. The vector $ai + bj + ck$ is called the direction vector of the line. Taking any point (x, y, z) on the line:
  (i) give the relation between line and its direction vector;
 (ii) deduce the equation of the straight line in analytical form.

2. Find a unit vector in the direction of the line

$$\frac{x-2}{2} = \frac{y-3}{3} = \frac{z+2}{6}$$

3. The line $\dfrac{x-2}{2} = \dfrac{y-3}{3} = \dfrac{z+2}{6}$ is parallel to the line

$$\frac{x-a}{d} = \frac{y-b}{e} = \frac{z-c}{f}$$

which passes through the point $(4, 4, 4)$. Deduce the values of a, b, c, d, e, f.

4. Show how to obtain the equation

$$\frac{x}{2} = \frac{y}{3} = \frac{z+8}{6}$$

from $\dfrac{x-2}{2} = \dfrac{y-3}{3} = \dfrac{z+2}{6}$

Do they represent the same straight line? Test by taking three points.

5. Find the equation of the line L joining the points $(3, 4, 7)$ and $0, 1, 3$. Give its direction vector and find a unit vector in the same direction. What is the length of the projection of the line joining $(5, 5, 5)$ to $(7, 6, 7)$ on the line L?

The normal unit vector to a plane

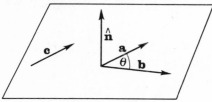

Fig. 107

A line drawn perpendicular to a plane is perpendicular to any straight line in that plane.

In vector terms this means that a vector in a plane is at right angles to the vector perpendicular to the plane.

In the diagram above **a**, **b** and **c** are three vectors in the plane indicated and **n̂** is the unit vector *normal* to the plane.

The scalar or dot product of two vectors is zero when the vectors are at right angles, hence

a . n̂ = 0

b . n̂ = 0

and

c . n̂ = 0

But since **n̂** is normal to the plane containing **a** and **b** then

b × **a** = |**b**| |**a**| sin θ**n̂**

$\qquad$ = k**n̂**

Since **c** is perpendicular to k**n̂** then

c . kn̂ = 0 ⇒ **c . (b × a)** = 0

We now use this to find the equation of a plane specified by three given points.

To find the equation of a plane
A plane is determined by three given points A (1, 2, 3), B (3, 3, 4) and C (4, 6, 6), find its equation.

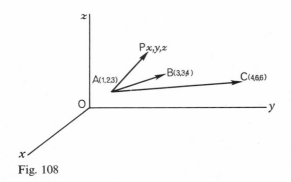

Fig. 108

119

Let P(x, y, z) be a point on this plane

Vector $\mathbf{AB} = 2\mathbf{i} + 1\mathbf{j} + 1\mathbf{k}$

Vector $\mathbf{AC} = 3\mathbf{i} + 4\mathbf{j} + 3\mathbf{k}$

$$\mathbf{AC} \times \mathbf{AB} = \begin{vmatrix} \mathbf{i} & \mathbf{j} & \mathbf{k} \\ 3 & 4 & 3 \\ 2 & 1 & 1 \end{vmatrix} = 1\mathbf{i} + 3\mathbf{j} - 5\mathbf{k} = \mathbf{n}$$

$\Rightarrow$ A vector perpendicular to plane of $\mathbf{AC}$ and $\mathbf{AB}$ is the vector $\mathbf{n} = 1\mathbf{i} + 3\mathbf{j} - 5\mathbf{k}$

For P to lie in this plane $\mathbf{AP}$ must be at right angles to vector $\mathbf{n}$ i.e. $\mathbf{n} \cdot \mathbf{AP} = 0$

$(1\mathbf{i} + 3\mathbf{j} - 5\mathbf{k}) \cdot \{(x-1)\mathbf{i} + (y-2)\mathbf{j} + (z-3)\mathbf{k}\} = 0$

$\Rightarrow x - 1 + 3(y-2) - 5(z-3) = 0$

$\Rightarrow x - 1 + 3y - 6 - 5z + 15 = 0$

$\Rightarrow x + 3y - 5z + 8 = 0$

The equation gives the relation between x, y and z for any point P which lies in the plane determined by the three points A, B and C.

The vector $\mathbf{n}$ normal to the plane ABC has a magnitude

$$|\mathbf{n}| = \sqrt{(+1)^2 + (+3)^2 + (-5)^2}$$

$$= \sqrt{35}$$

Denoting the *unit* normal vector by $\hat{\mathbf{n}}$ means

$\mathbf{n} = |\mathbf{n}|\hat{\mathbf{n}}$

Hence

$$\hat{\mathbf{n}} = \frac{\mathbf{n}}{|\mathbf{n}|} = \frac{1\mathbf{i} + 3\mathbf{j} - 5\mathbf{k}}{\sqrt{35}}$$

$$= \frac{1\mathbf{i}}{\sqrt{35}} + \frac{3\mathbf{j}}{\sqrt{35}} - \frac{5\mathbf{k}}{\sqrt{35}}$$

or $\dfrac{\sqrt{35}\mathbf{i}}{35} + \dfrac{3\sqrt{35}\mathbf{j}}{35} - \dfrac{5\sqrt{35}\mathbf{k}}{35}$

To find the area of △ABC in the given plane

From previous work $|\mathbf{AC} \times \mathbf{AB}|$ = Area of vector parallelogram

$$= 2(\text{Area of } \triangle ABC)$$

$$= \sqrt{35} \text{ (magnitude of the vector } \mathbf{n})$$

$$\text{Area of } \triangle ABC = \frac{\sqrt{35}}{2} \text{ sq units}$$

Equation of a plane in vector form

To find the vector equation of the plane perpendicular to a given vector **a** and passing through a given point B.

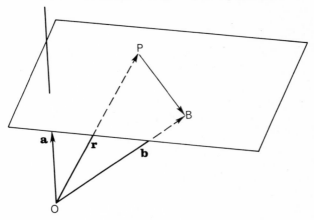

Fig. 109

In Figure 109 let **a** be the given vector, P any point in the required plane and B the given point. Calling the position vectors of B and P, **b** and **r** respectively from a convenient origin O, then **PB** = (**b**−**r**) and since **PB** is perpendicular to **a** the scalar product must be zero

$$\Rightarrow (\mathbf{b}-\mathbf{r}) \cdot \mathbf{a} = 0$$

or

$$\mathbf{r} \cdot \mathbf{a} = \mathbf{b} \cdot \mathbf{a}$$

This is the required vector equation, with **r** the variable vector, **a** and **b** are constant vectors for this plane.

Since **a** is a free vector perpendicular to the required plane it is convenient to let its direction pass through the origin of the

121

rectangular axes and then choose this origin for the position vectors. The vector form of the equation is easily reduced to the analytical form.

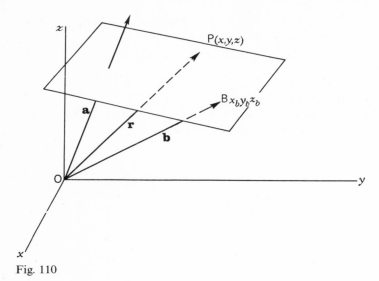

Fig. 110

Let $\mathbf{a} = l\mathbf{i} + m\mathbf{j} + n\mathbf{k}$

$\mathbf{OP} = \mathbf{r} = x\mathbf{i} + y\mathbf{j} + z\mathbf{k}$

$\mathbf{OB} = \mathbf{b} = x_b\mathbf{i} + y_b\mathbf{j} + z_b\mathbf{k}$

The vector form of the equation of the plane $\mathbf{r} \cdot \mathbf{a} = \mathbf{b} \cdot \mathbf{a}$ then becomes:

$(x\mathbf{i} + y\mathbf{j} + z\mathbf{k}) \cdot (l\mathbf{i} + m\mathbf{j} + n\mathbf{k}) = (x_b\mathbf{i} + y_b\mathbf{j} + z_b\mathbf{k}) \cdot (l\mathbf{i} + m\mathbf{j} + n\mathbf{k})$

$\Rightarrow lx + my + nz = lx_b + my_b + nz_b$

Since l, m, n, x_b, y_b, z_b are constants for the plane then

$lx_b + my_b + nz_b = \text{constant} = d$

Hence the equation of the plane in analytical form is

$lx + my + nz - d = 0$

It is important to notice that the coefficients of x, y, and z are the components of the vector perpendicular to the plane and the value of d merely depends on the position of the given point B.

Example

The equation of a plane is $2x+3y+2z-4 = 0$. Find (i) the vector normal to the plane and (ii) the unit vector normal to the plane.

(i) The vector normal to the plane is $2\mathbf{i}+3\mathbf{j}+2\mathbf{k}$ from the work above. Call this normal vector $\mathbf{n}$

$$|\mathbf{n}| = \sqrt{2^2+3^2+2^2} = \sqrt{17}$$

$\mathbf{n} = |\mathbf{n}|.\hat{\mathbf{n}}$ where $\hat{\mathbf{n}}$ is the unit normal vector

(ii) Unit normal vector

$$\hat{\mathbf{n}} = \frac{\mathbf{n}}{|\mathbf{n}|} = \frac{1}{\sqrt{17}}(2\mathbf{i}+3\mathbf{j}+2\mathbf{k}) = \frac{2}{\sqrt{17}}\mathbf{i}+\frac{3}{\sqrt{17}}\mathbf{j}+\frac{2}{\sqrt{17}}\mathbf{k}$$

(iii) Find the perpendicular distance from O to the plane.

The point $P(3, -2, 2)$ lies in the given plane.

$$\mathbf{OP} = 3\mathbf{i}-2\mathbf{j}+2\mathbf{k}$$

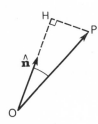

Fig. 111a

The projection of **OP** on the normal unit vector is the perpendicular distance required because

$$\mathbf{OP}.\hat{\mathbf{n}} = |\mathbf{OP}|\,|\hat{\mathbf{n}}|\cos\theta \text{ where angle } \mathbf{HOP} = \theta$$

$$= |\mathbf{OP}|\cos\theta = \mathbf{OH} = h$$

$$h = (3\mathbf{i}-2\mathbf{j}+2\mathbf{k}).\frac{1}{\sqrt{17}}(2\mathbf{i}+3\mathbf{j}+2\mathbf{k})$$

$$= \frac{1}{\sqrt{17}}(6-6+4) = \frac{4}{\sqrt{17}} = \frac{4\sqrt{17}}{17}$$

Exercise 11b

1. Find a direction vector for the straight line

$$\frac{x-2}{3} = \frac{y-4}{6} = \frac{z-5}{9}$$

i.e., a vector having the same direction.

2. Find a unit vector in the direction of the line

$$\frac{x+1}{1} = \frac{y+2}{2} = \frac{3+4}{3}$$

3. The point $(3, 6, 8)$ lies on the line

$$\frac{x-2}{3} = \frac{y-4}{6} = \frac{z-5}{9}$$

Does it lie on line

$$\frac{x+1}{1} = \frac{y+2}{2} = \frac{z+4}{3}?$$

Test another point. What do you conclude about the two lines

$$\frac{x-2}{3} = \frac{y-4}{6} = \frac{z-5}{9}$$

and

$$\frac{x+1}{2} = \frac{y+2}{2} = \frac{z+4}{3}?$$

4. What are the direction cosines of the line

$$\frac{x-2}{1} = \frac{y-4}{2} = \frac{z-5}{3}?$$

5. What is the direction vector of the line

$$\frac{x-a}{p} = \frac{y-b}{q} = \frac{z-c}{r}?$$

6. What can you say about the lines

$$\frac{x+1}{1} = \frac{y+2}{2} = \frac{z+4}{3}$$

and

$$\frac{x-1}{1} = \frac{y+3}{2} = \frac{z-2}{3}?$$

7. Find the perpendicular distance from the point $(3, 4, 5)$ to the line

$$\frac{x+1}{1} = \frac{y+2}{2} = \frac{z+4}{3}$$

8. Find a vector perpendicular to the two lines

$$\frac{x+1}{1} = \frac{y+2}{2} = \frac{z+4}{3}$$

and

$$\frac{x-2}{3} = \frac{y-4}{2} = \frac{z-5}{4}$$

9. (i) Do the lines

$$\frac{x+1}{1} = \frac{y+2}{2} = \frac{z+4}{3}$$

and

$$\frac{x-2}{3} = \frac{y-4}{2} = \frac{z-5}{4}$$

intersect?

(ii) What kind of lines are these called?

(iii) What is the perpendicular distance between them?

10. What is the locus of vectors perpendicular to the line

$$\frac{x}{2} = \frac{y}{3} = \frac{z+8}{6}$$

and passing through the point $(4, 4, 4)$?

Intersecting planes

Unless two planes are parallel they intersect in a straight line. The angle between the two intersecting planes is called a DIHEDRAL angle and is defined as the angle between the normal vectors to the two planes. That this is true is clear from the figure, if we consider the outward drawn normals for each plane.

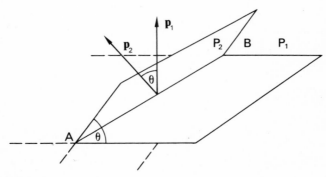

Fig. 111b

If the equation of plane P_1 is $a_1 x + b_1 y + c_1 z + d_1 = 0$ and of plane P_2 is $a_2 x + b_2 y + c_2 z + d_2 = 0$, then the normal vector $\mathbf{p}_1$ of plane P_1 is given by $\mathbf{p}_1 = a_1 \mathbf{i} + b_1 \mathbf{j} + c_1 \mathbf{k}$ and the normal vector to plane P_2 is $\mathbf{p}_2 = a_2 \mathbf{i} + b_2 \mathbf{j} + c_2 \mathbf{k}$.

Using the Scalar Product

$$\mathbf{p}_1 \cdot \mathbf{p}_2 = |\mathbf{p}_1||\mathbf{p}_2| \cdot \cos \theta$$

hence $\theta = \cos^{-1} \dfrac{\mathbf{p}_1 \cdot \mathbf{p}_2}{|\mathbf{p}_1||\mathbf{p}_2|}$

$$= \cos^{-1} \frac{a_1 a_2 + b_1 b_2 + c_1 c_2}{\sqrt{(a_1^2 + b_1^2 + c_1^2)(a_2^2 + b_2^2 + c_2^2)}}$$

For the planes to be parallel $\theta = 0$, i.e., $\cos \theta = 1$, and this condition is satisfied if

$$\frac{a_1}{a_2} = \frac{b_1}{b_2} = \frac{c_1}{c_2} = K \text{ (some constant)}$$

The line of intersection of two planes
Referring again to Figure 111b, the line of intersection AB of the two planes lies in plane P_1 is therefore perpendicular to the normal vector $\mathbf{p}_1$, also AB lies in plane P_2 and is also perpendicular to the normal $\mathbf{p}_2$. Since the line AB is perpendicular to both the normal vectors $\mathbf{p}_1$ and $\mathbf{p}_2$, the direction vector of AB is given by the Vector Product of $\mathbf{p}_1$ and $\mathbf{p}_2$.

$$\mathbf{p}_1 \times \mathbf{p}_2 = \begin{vmatrix} \mathbf{i} & \mathbf{j} & \mathbf{k} \\ a_1 & b_1 & c_1 \\ a_2 & b_2 & c_2 \end{vmatrix}$$

$$= d\mathbf{i} + e\mathbf{j} + f\mathbf{k}$$

where $d = b_1 c_2 - b_2 c_1$

$\qquad e = a_2 c_1 - a_1 c_2$

and $\quad f = a_1 b_2 - a_2 b_1$

The equation of the line AB can be written

$$\frac{x-s}{d} = \frac{y-t}{e} = \frac{z-u}{f} = \lambda$$

where λ is an arbitrary constant. Since d, e, and f are known it is only necessary to find one point on the line to determine the constants s, t, and u. This one point can be found easily by determining where the planes P_1 and P_2 cut the plane $z = 0$ simultaneously, i.e., by solving

$$\left. \begin{array}{l} a_1 x + b_1 y + d_1 = 0 \\ a_2 x + b_2 y + d_2 = 0 \end{array} \right\}$$

and

simultaneously to find x and y where $z = 0$. Then by giving λ an arbitrary value, say 2, s, t, and u can be found, and the equation of AB is then fully known. The equation of the line AB so found is not unique since it depends of the value given to λ, but the line AB that these equations represent is certainly unique. As an example, consider the equation

$$\frac{x-3}{2} = \frac{y-3}{4} = \frac{z-4}{5} \, (= \lambda \text{ say})$$

Putting $\lambda = 1$ gives $x - 3 = 2$ or $x = 5$

$\qquad\qquad\qquad\qquad y - 3 = 4$ or $y = 7$

and $\qquad\qquad\qquad z - 4 = 5$ or $z = 9$

hence the point (5, 7, 9) lies on the line. Putting λ equal to other values gives other points on the same line. Suppose we add 2 to each fraction thus

$$\frac{x-3}{2}+2 = \frac{y-3}{4}+2 = \frac{z-4}{5}+2 = (\lambda \text{ say})$$

the equation then becomes

$$\frac{x+1}{2} = \frac{y+5}{4} = \frac{z+6}{5}$$

and the values $x = 5$, $y = 7$, and $z = 9$ still satisfy the new equation. The new equation still represents the same line, although its form has changed, but the direction vector of both forms remains unaltered.

Exercise 12.

1. If the position vectors of A, B and C of the plane triangle ABC are $\mathbf{r}_1$, $\mathbf{r}_2$ and $\mathbf{r}_3$, find the area of the triangle in terms of $\mathbf{r}_1$, $\mathbf{r}_2$, and $\mathbf{r}_3$.

2. If the position vectors of the points A, B, C are $\mathbf{r}_1$, $\mathbf{r}_2$ and $\mathbf{r}_3$, what is the condition for collinearity of A, B and C.

3. What is the condition for sides AB and BC to be at right angles in Qu. 1.

4. Find the equation of the plane determined by points A(2, 3, 3), B(3, 4, 6) and C(5, 7, 8).

5. Find the unit vector normal to the plane in Qu. 4.

6. Find the area of $\triangle$ABC in Qu. 4.

7. Find the equation of the line joining A(2, 3, 3) and B(3, 4, 6) in vector form.

8. Find the equation of the line joining A(2, 3, 3) and C(5, 7, 8) in analytical form.

9. Find the distance from the point P(1, 3, 6) to the plane determined by points A(2, 3, 3), B(3, 4, 6) and C(5, 7, 8).

10. If the position vectors of the vertices of the quadrilateral A, B, C, D, are **a**, **b**, **c**, **d** respectively, show that the area of the quadrilateral ABCD is

$\frac{1}{2}|(\mathbf{a} \times \mathbf{b}) + (\mathbf{b} \times \mathbf{c}) + (\mathbf{c} \times \mathbf{d}) + (\mathbf{d} \times \mathbf{a})|$

11. The vector $\mathbf{r} = a\mathbf{i} + b\mathbf{j} + c\mathbf{k}$ is perpendicular to the given plane passing through the origin (0, 0). Prove that the equation of this plane is $ax + by + cz = 0$. (*Hint:* take any point P(x, y, z) on the plane and find the scalar product of **r** and **OP**.)

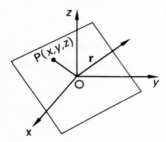

Fig. 111c

12. OABC is a tetrahedron. The area of triangle OAB $= \frac{1}{2}|\mathbf{a} \times \mathbf{b}|$. Let $\mathbf{S}_1$ be a vector normal to the plane of triangle OAB and acting outwards, with magnitude equal to $\frac{1}{2}|\mathbf{a} \times \mathbf{b}|$, then $\mathbf{S}_1 = \frac{1}{2}(\mathbf{a} \times \mathbf{b})$ and forming a right-handed triple with **a** and **b**. Similarly, $\mathbf{S}_2, \mathbf{S}_3, \mathbf{S}_4$, are vectors equal in magnitude to the areas of the other faces as shown and acting outwards. Show that the sum of the area vectors is zero, i.e., $\mathbf{S}_1 + \mathbf{S}_2 + \mathbf{S}_3 + \mathbf{S}_4 = \mathbf{0}$.

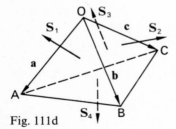

Fig. 111d

13. $(x'y'z')$ is a point on the sphere of radius a with its centre at the origin. What is the equation of the plane tangential to the sphere at the point (x', y', z')?

128

14. A cylinder has the equation $(x-2)^2+(y-3)^2 = 25$. Find the equation of the straight line which passes through the point $(6, 6, 8)$ on the surface of the cylinder, lies in the tangential plane to the cylinder and cuts the x-axis.

15. Find a unit vector perpendicular to the plane

$$3x+2y+6z-20 = 0.$$

Show that the point $(2, 4, 1)$ lies in the plane and find the perpendicular distance from the point $(4, 6, 4)$ to the plane. (*Hint:* the scalar product of a vector with a unit vector gives the component of the vector in the direction of the unit vector.)

16. What is the locus of vectors all perpendicular to the line

$$\frac{x-2}{2} = \frac{y-3}{3} = \frac{z+2}{6}?$$

17. Find the line of intersection of the two planes

$$3x+2y+6z-20 = 0$$

and

$$x-2y+3z-10 = 0$$

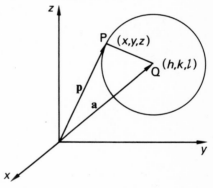

Fig. 111e

18. In the sphere shown in Fig. 111e,

$$\mathbf{r} = \mathbf{p}-\mathbf{a} = (x-h)\mathbf{i}+(y-k)\mathbf{j}+(z-l)\mathbf{k}$$

By using the scalar product $\mathbf{r} \cdot \mathbf{r}$, establish the equation of the sphere in cartesian coordinates.

19. Treating $\mathbf{r}$ as a vector normal to the tangent plane to the sphere of Qu. 18 at point (a, b, c), find the equation of the tangent plane in cartesian coordinates.

20. If the centre of the sphere is now at the origin find the radius vector $(\mathbf{v})$ which makes an angle of $45°$ with the positive directions of the x and y axes.

21. A sphere of radius 5 units has its centre at $Q(7, 6, 5)$. What is the equation of the plane which is tangential to the sphere and makes an angle of $135°$ with the positive direction of the y-axis.

129

Chapter 9
Triple products

Triple scalar products

On page 110 in order to find the equation of a plane in analytical form we first found a cross or vector product of two vectors $(\mathbf{AC} \times \mathbf{AB})$, and then this vector was used with a third vector $\mathbf{AP}$ to find a scalar product.

In all this amounted to

$$\mathbf{AP} . (\mathbf{AC} \times \mathbf{AB})$$

Such a product is called a TRIPLE SCALAR PRODUCT. Consider

$$\mathbf{a} . \mathbf{b} \times \mathbf{c}$$

It is not necessary to use brackets to specify the order of operations because $\mathbf{a} . \mathbf{b}$ is a scalar quantity and the cross product or vector product of a scalar with a vector has no meaning. Such triple products as $\mathbf{a} . \mathbf{b} \times \mathbf{c}$ arise in many problems and the calculation can be shortened by using the determinant method.

Let

$$\mathbf{a} = x_a\mathbf{i} + y_a\mathbf{j} + z_a\mathbf{k}$$

$$\mathbf{b} = x_b\mathbf{i} + y_b\mathbf{j} + z_b\mathbf{k}$$

$$\mathbf{c} = x_c\mathbf{i} + y_c\mathbf{j} + z_c\mathbf{k}$$

where x_a, y_a and z_a are the components of vector $\mathbf{a}$ in the base directions, similarly for vectors $\mathbf{b}$ and $\mathbf{c}$.

$$\mathbf{b} \times \mathbf{c} = \begin{vmatrix} \mathbf{i} & \mathbf{j} & \mathbf{k} \\ x_b & y_b & z_b \\ x_c & y_c & z_c \end{vmatrix} = \begin{vmatrix} y_b & z_b \\ y_c & z_c \end{vmatrix} \mathbf{i} - \begin{vmatrix} x_b & z_b \\ x_c & z_c \end{vmatrix} \mathbf{j} + \begin{vmatrix} x_b & y_b \\ x_c & y_c \end{vmatrix} \mathbf{k}$$

Let this be equal to $X_1\mathbf{i} - Y_1\mathbf{j} + Z_1\mathbf{k}$ where X_1, Y_1 and Z_1 stand for the 2nd order determinants.

Then $\mathbf{a} . \mathbf{b} \times \mathbf{c}$ is the inner product or scalar product of $x_a\mathbf{i} + y_a\mathbf{j} + z_a\mathbf{k}$ and $X_1\mathbf{i} - Y_1\mathbf{j} + Z_1\mathbf{k}$

$$\mathbf{a} . \mathbf{b} \times \mathbf{c} = x_aX_1 - y_aY_1 + z_aZ_1 \text{ (since all other products are zero)}$$

$$= x_a \begin{vmatrix} y_b & z_b \\ y_c & z_c \end{vmatrix} - y_a \begin{vmatrix} x_b & z_b \\ x_c & z_c \end{vmatrix} + z_a \begin{vmatrix} x_b & y_b \\ x_c & y_c \end{vmatrix}$$

But this is the expansion of the determinant

$$\begin{vmatrix} x_a & y_a & z_a \\ x_b & y_b & z_b \\ x_c & y_c & z_c \end{vmatrix}$$

$$\Rightarrow \mathbf{a} \cdot \mathbf{b} \times \mathbf{c} = \begin{vmatrix} x_a & y_a & z_a \\ x_b & y_b & z_b \\ x_c & y_c & z_c \end{vmatrix}$$

The determinant form is easy to memorise.

Since the interchange of two rows of a determinant alters the sign from positive to negative or vice-versa, two such row interchanges will leave the sign unchanged, thus:

$$\begin{vmatrix} x_a & y_a & z_a \\ x_b & y_b & z_b \\ x_c & y_c & z_c \end{vmatrix} = - \begin{vmatrix} x_b & y_b & z_b \\ x_a & y_a & z_a \\ x_c & y_c & z_c \end{vmatrix} = \begin{vmatrix} x_b & y_b & z_b \\ x_c & y_c & z_c \\ x_a & y_a & z_a \end{vmatrix} = \mathbf{b} \cdot (\mathbf{c} \times \mathbf{a})$$

Also

$$\begin{vmatrix} x_a & y_a & z_a \\ x_b & y_b & z_b \\ x_c & y_c & z_c \end{vmatrix} = - \begin{vmatrix} x_c & y_c & z_c \\ x_b & y_b & z_b \\ x_a & y_a & z_a \end{vmatrix} = \begin{vmatrix} x_c & y_c & z_c \\ x_a & y_a & z_a \\ x_b & y_b & z_b \end{vmatrix} = \mathbf{c} \cdot (\mathbf{a} \times \mathbf{b})$$

$$\Rightarrow \mathbf{a} \cdot \mathbf{b} \times \mathbf{c} = \mathbf{b} \cdot \mathbf{c} \times \mathbf{a} = \mathbf{c} \cdot \mathbf{a} \times \mathbf{b}$$

and since scalar products are commutative

$$= \mathbf{b} \times \mathbf{c} \cdot \mathbf{a} = \mathbf{c} \times \mathbf{a} \cdot \mathbf{b} = \mathbf{a} \times \mathbf{b} \cdot \mathbf{c}$$

The triple scalar product as a volume

The three vectors $\mathbf{a}$, $\mathbf{b}$ and $\mathbf{c}$ form the sides of the parallelepiped shown.

The vector product $\mathbf{b} \times \mathbf{c} = $ (Area of Base parallelogram)$\hat{\mathbf{n}}$ where $\hat{\mathbf{n}}$ is a unit vector normal to the plane of the base.

The scalar product of $\mathbf{a}$ with this unit vector $\hat{\mathbf{n}}$ has been shown previously to be equal to the length of the projection of vector $\mathbf{a}$ on the direction of the normal vector $\hat{\mathbf{n}}$ i.e. $\mathbf{a} \cdot \hat{\mathbf{n}} = AH$.

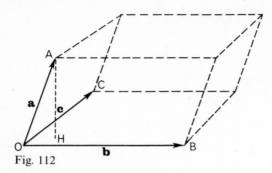

Fig. 112

It follows that

$$\mathbf{a} \cdot \mathbf{b} \times \mathbf{c} = \mathbf{a} \cdot (\text{Area of Base}) \hat{\mathbf{n}}$$
$$= (\text{Area of Base}) \mathbf{a} \cdot \hat{\mathbf{n}}$$
$$= (\text{Area of Base}) \cdot \text{AH}$$
$$= \text{Volume of parallelepiped}$$

Since $\mathbf{b} \cdot \mathbf{c} \times \mathbf{a} =$ Volume of same parallelepiped and $\mathbf{c} \cdot \mathbf{a} \times \mathbf{b} =$ Volume of same parallelepiped once more we show that $\mathbf{a} \cdot \mathbf{b} \times \mathbf{c} = \mathbf{b} \cdot \mathbf{c} \times \mathbf{a} = \mathbf{c} \cdot \mathbf{a} \times \mathbf{b}$.

Note on order of multiplication
Since

$$\mathbf{b} \times \mathbf{c} = -\mathbf{c} \times \mathbf{b}$$

Then

$$\mathbf{a} \cdot \mathbf{c} \times \mathbf{b} = -\mathbf{a} \cdot \mathbf{b} \times \mathbf{c}$$

$$= -\text{Volume of parallelepiped}$$

The *magnitude* of the triple scalar product is still equal to the volume of the parallelepiped.

Condition for three vectors to be coplanar

If the three vectors $\mathbf{a}$, $\mathbf{b}$ and $\mathbf{c}$ are coplanar then the volume of the vector parallelepiped must be zero, i.e.

$$\begin{vmatrix} x_a & y_a & z_a \\ x_b & y_b & z_b \\ x_c & y_c & z_c \end{vmatrix} = 0$$

This is the required condition.

132

The triple cross product

The vector product $\mathbf{a} \times (\mathbf{b} \times \mathbf{c})$ is called the triple vector product of the three vectors $\mathbf{a}$, $\mathbf{b}$ and $\mathbf{c}$.

Compared with the triple scalar product, the triple vector product has limited applications (mainly in spherical trigonometry).

It can be shown that

$$\mathbf{a} \times (\mathbf{b} \times \mathbf{c}) = (\mathbf{a} \cdot \mathbf{c})\mathbf{b} - (\mathbf{a} \cdot \mathbf{b})\mathbf{c}$$

A number of different methods are available, some more elegant than others, the most straightforward method is rather lengthy using the components of each vector:

$$\mathbf{a} \times (\mathbf{b} \times \mathbf{c}) = (x_a\mathbf{i} + y_a\mathbf{j} + z_a\mathbf{k}) \times \begin{vmatrix} \mathbf{i} & \mathbf{j} & \mathbf{k} \\ x_b & y_b & z_b \\ x_c & y_c & z_c \end{vmatrix}$$

which when multiplied and expressed in component form reduces to the same expression as

$(\mathbf{a} \cdot \mathbf{c})\mathbf{b} - (\mathbf{a} \cdot \mathbf{b})\mathbf{c}$ (also expressed in component form)

Exercise 13.

1. If $\mathbf{a} = 2\mathbf{i} + 2\mathbf{j} + 3\mathbf{k}$, $\mathbf{b} = 2\mathbf{i} + 3\mathbf{j} + 4\mathbf{k}$, $\mathbf{c} = 3\mathbf{i} + 4\mathbf{j} + 7\mathbf{k}$. Find $(\mathbf{a} \times \mathbf{b})$, $\mathbf{c} \cdot (\mathbf{a} \times \mathbf{b})$, $(\mathbf{a} \times \mathbf{c})$, $\mathbf{b} \cdot (\mathbf{a} \times \mathbf{c})$.
2. If the vector $\mathbf{p} = x\mathbf{i} + 3\mathbf{j} + 2\mathbf{z}$ is coplanar with vectors $\mathbf{a} = 2\mathbf{i} + 2\mathbf{j} + 3\mathbf{k}$ and $\mathbf{b} = 2\mathbf{i} + 3\mathbf{j} + 4\mathbf{k}$. Find the value of the component x.
3. By expanding the determinant on page 118 show that $\mathbf{a} \times (\mathbf{b} \times \mathbf{c}) = (\mathbf{a} \cdot \mathbf{c})\mathbf{b} - (\mathbf{a} \cdot \mathbf{b})\mathbf{c}$.
4. Prove that $\mathbf{a} \cdot (\mathbf{b} \times \mathbf{c}) = (\mathbf{a} \times \mathbf{b}) \cdot \mathbf{c}$.
5. Prove that $(\mathbf{a} \times \mathbf{b}) \times \mathbf{c} = \mathbf{b}(\mathbf{a} \cdot \mathbf{c}) - \mathbf{a}(\mathbf{b} \cdot \mathbf{c})$.
6. Prove that $\mathbf{a} \times (\mathbf{b} \times \mathbf{c}) + \mathbf{b} \times (\mathbf{c} \times \mathbf{a}) + \mathbf{c} \times (\mathbf{a} \times \mathbf{b}) = 0$.
7. If $\mathbf{a}$, $\mathbf{b}$, $\mathbf{c}$ and $\mathbf{d}$ are coplanar, prove that
   $(\mathbf{a} \times \mathbf{b}) \times (\mathbf{c} \times \mathbf{d}) = 0$
8. Use the result of Qu. 4 to show that $\mathbf{a} \cdot (\mathbf{a} \times \mathbf{c}) = 0$.
9. Using the result of Qu. 4, or otherwise, show that $(\mathbf{a} \times \mathbf{b}) \cdot (\mathbf{c} \times \mathbf{d}) = (\mathbf{a} \cdot \mathbf{c})(\mathbf{b} \cdot \mathbf{d}) - (\mathbf{a} \cdot \mathbf{d})(\mathbf{b} \cdot \mathbf{c})$.
10. Plane P_1 contains vectors $\mathbf{a}$ and $\mathbf{b}$, plane P_2 contains vectors $\mathbf{c}$ and $\mathbf{d}$:
    (i) Give vectors perpendicular to the planes.
    (ii) What is the condition for the planes P_1 and P_2 to be parallel?

(iii) What is the condition for the planes P_1 and P_2 to be perpendicular?

11. **a**, **b**, and **c** are non-equal vectors and **a** and **b** are coplanar. If **d** = **a**×**b** and **e** = **d**×**c**, what can you conclude about vector **e**? If **e** = **0**, can you deduce any further information?

Chapter 10
Vectors and mechanics

Many of the concepts of mechanics are really vector quantities and their uses become more meaningful when they are regarded vectorially. Many of the laws and definitions of mechanics can be rephrased in terms of vectors, and the concepts of scalar product of two vectors and the vector product of two vectors serve to extend the application of vectorial methods.

The triangle of forces as a method of compounding or *adding* forces was first stated explicitly by Simon Stevin of Bruges in *1584*. He used a straight line to represent a force in magnitude, direction and sense. Later came the extension of the triangle method for the compounding of velocities where a straight line is used to represent a velocity in magnitude, direction and sense. So that long before the word vector had been invented by Sir W. R. Hamilton in 1854 to represent such quantities, mathematicians and engineers had been developing the concept of such a quantity and its representation geometrically by a line segment. James Watt in 1784 introduced the concept of *work done* as a scalar quantity resulting from the product of two vector quantities but it was not until 1884 that Professor W. Gibbs formulated the idea of the scalar product of two vectors in its general form as was stated in Chapter 5. In the early part of the nineteenth century engineers and physicists were using *vector fields* and vector products but limited them to electromagnetism and they did not have the advantages of vector algebra to clarify and generalise their results. The study of kinematics, mechanics, hydrodynamics, electricity and magnetism require a wide application of vector methods, and in particular the use of calculus methods. The subject is too deep and vast for a comprehensive treatment, so first we shall show that the methods of the calculus can be applied to vectors and we shall use the result we obtain to illustrate some of the elementary but widely used concepts of mechanics.

Position vectors and the calculus
In number algebra, if y is a function of x then the Derived Function of y (or Derivative) is defined as

$$y' \left(\text{or } \frac{dy}{dx} \right) = \lim_{\delta x \to 0} \frac{f(x+\delta x) - f(x)}{\delta x}$$

We can transfer the methods of the calculus to vector algebra and obtain meaningful results.

Suppose a particle has a position vector **r** at a time t and if the displacement is a function of the time then we can find a meaning for $d\mathbf{r}/dt$. Let the position vector **r** become $\mathbf{r}+\delta\mathbf{r}$ after an interval of time δt

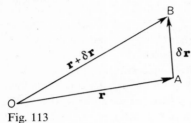

Fig. 113

The vector **AB** is the difference of **OB** and **OA**

$$\mathbf{AB} = \mathbf{r}+\delta\mathbf{r}-\mathbf{r} = \delta\mathbf{r}$$

Using the definition above we get

$$\frac{d\mathbf{r}}{dt} = \lim_{\delta t\to 0}\frac{(\mathbf{r}+\delta\mathbf{r})-\mathbf{r}}{\delta t} = \lim_{\delta t\to 0}\frac{\delta\mathbf{r}}{\delta t}$$

Since division in vector algebra is not defined, then we must regard $\dfrac{\delta\mathbf{r}}{\delta t}$ as $\delta\mathbf{r}$ times $\left(\dfrac{1}{\delta t}\right)$

i.e. the vector $\delta\mathbf{r}$ has been *multiplied* by the scalar $\left(\dfrac{1}{\delta t}\right)$

hence

$\left(\dfrac{\delta\mathbf{r}}{\delta t}\right)$ is a vector

$\delta\mathbf{r}/\delta t$ has the direction of $\delta\mathbf{r}$ i.e. **AB**

As $\delta t \to 0$ then point B approaches point A more and more closely until in the limiting position $\delta\mathbf{r}$ will be at right angles to the direction of vector **r**, if **r** is constant but not otherwise.

The rate of change of position with respect to time and *relative to the point O* is the velocity of the end point of vector **r** relative to O, at time t.

Hence $\quad \lim_{\delta t\to 0}\dfrac{\delta\mathbf{r}}{\delta t} = \mathbf{v}$

i.e. $\dfrac{d\mathbf{r}}{dt} = \mathbf{v}$

Similarly, $d\mathbf{v}/dt$ is the acceleration of the end point of vector **v** relative to the origin O and

$$\frac{d\mathbf{v}}{dt} = \frac{d^2\mathbf{r}}{dt^2}$$

We have shown that the process of differentiation applied to vector algebra gives meaningful results but many results have to be established, none can be assumed. After the next exercises in which some fundamental results are established from first principles, some important and far from obvious relations are derived from the differentiation of a unit vector.

Exercise 14.

1. Show that if **a** and **b** are vector functions of t then

$$\frac{d}{dt}(\mathbf{a}+\mathbf{b}) = \frac{d\mathbf{a}}{dt}+\frac{d\mathbf{b}}{dt}$$

2. If m is a scalar function of t and **a** is a vector function of t also, show that

$$\frac{d}{dt}(m\mathbf{a}) = \frac{dm}{dt}\mathbf{a}+m\frac{d\mathbf{a}}{dt}$$

3. If $\mathbf{r} = x\mathbf{i}+y\mathbf{j}+z\mathbf{k}$, show from first principles that

$$\frac{d\mathbf{r}}{dt} = \frac{dx}{dt}\mathbf{i}+\frac{dy}{dt}\mathbf{j}+\frac{dz}{dt}\mathbf{k}$$

4. Show that

$$\frac{d}{dt}(\mathbf{a}\cdot\mathbf{b}) = \mathbf{a}\cdot\frac{d\mathbf{b}}{dt}+\mathbf{b}\cdot\frac{d\mathbf{a}}{dt}$$

5. Show that

$$\frac{d}{dt}(\mathbf{a}\times\mathbf{b}) = \mathbf{a}\times\frac{d\mathbf{b}}{dt}+\frac{d\mathbf{a}}{dt}\times\mathbf{b}$$

6. Using the results established in questions 4 and 5 show that

$$\frac{d}{dt}(\mathbf{a}\cdot\mathbf{b}\times\mathbf{c}) = \mathbf{a}\cdot\mathbf{b}\times\frac{d\mathbf{c}}{dt}+\mathbf{a}\cdot\frac{d\mathbf{b}}{dt}\times\mathbf{c}+\frac{d\mathbf{a}}{dt}\cdot\mathbf{b}\times\mathbf{c}$$

7. If $r = \mathbf{a}\cos nt+\mathbf{b}\sin nt$ find $d\mathbf{r}/dt$ and $d^2\mathbf{r}/dt^2$ and show that $\ddot{\mathbf{r}}+n^2\mathbf{r} = 0$. ($\ddot{\mathbf{r}}$ represents $d^2\mathbf{r}/dt^2$).

8. If $\mathbf{r} = \sin t\mathbf{i}+\cos t\mathbf{j}+t\mathbf{k}$ find $d\mathbf{r}/dt$ and $d^2\mathbf{r}/dt^2$ and calculate $|d\mathbf{r}/dt|$ and $|d^2\mathbf{r}/dt^2|$.

9. The position vector of point P is $\mathbf{r} = \mathbf{a}\cos\omega t+\mathbf{b}\sin\omega t$, where **a** and **b** are constant vectors and ω is a constant, t is a variable (time).

Find the acceleration and velocity of point P. If $|\mathbf{a}| = |\mathbf{b}|$ and $\mathbf{a}\cdot\mathbf{b} = 0$, give the directions of **v** and $\dot{\mathbf{v}}$.

10. The position vector of a point P is given by $\mathbf{r} = k\cos\omega t\,\mathbf{a}$,

where **a** is a constant vector. Give expressions for the velocity and acceleration of P. Identify the motion.

11. A point moves with constant velocity. At time t_1 its position vector is $\mathbf{r}_1$ and at time t_2 its position vector is $\mathbf{r}_2$. What is its velocity vector?

12. At a time t the position vector of a body is

$$\mathbf{r} = at^2\mathbf{i} + bt\mathbf{j} + 3\mathbf{k}$$

What is its speed at time $t = 3$.

13. A particle P is at point B at time $t = 0$ and is moving with a constant speed s in the direction of a constant vector **a**. It is observed from two different points 0 and 0'. If after time t the position vectors of P are **r** and **r**' from 0 and 0', find these position vectors in terms of s, **b**, and t. Find $\dot{\mathbf{r}}$ and $\dot{\mathbf{r}}'$, and after comparing them state your conclusions.

(*Note:* $d(\mathbf{r})/dt = \dot{\mathbf{r}} = \mathbf{v}$)

Rotating Unit-Vector

In the application of vector algebra to dynamics the properties of the unit vector are most important, in particular the property of the rotating unit vector. The magnitude of the rotating unit vector is constant although its direction is continuously altering. In Figure 114 the unit vector **u** is shown initially in the position OP and then after an interval of time δt in the new position OQ. The use of the delta notation implies that the increment of time is exceedingly small and the angle will also be of the same order in size.

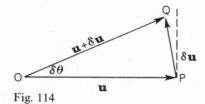

Fig. 114

If the unit vector is rotating with constant angular velocity, then $\delta\theta/\delta t = \omega$.

In this case we can show that **u** and $d\mathbf{u}/dt$ are perpendicular to each other. Since **u** is constant in magnitude and we have shown earlier that $|\mathbf{u}|^2 = \mathbf{u} \cdot \mathbf{u}$ (the scalar product) it follows that $\mathbf{u} \cdot \mathbf{u} =$ constant $= 1$ unit

$$\Rightarrow \frac{d}{dt}(\mathbf{u} \cdot \mathbf{u}) = 0$$

and using the result of Qu. 4, Ex. 14., we have

$$2\mathbf{u} \cdot \frac{d\mathbf{u}}{dt} = 0$$

Now since $\mathbf{u} \cdot (d\mathbf{u}/dt) = 0$ it has been shown that this is the condition for the two vectors $\mathbf{u}$ and $d\mathbf{u}/dt$ to be perpendicular to each other since both $\mathbf{u}$ and $d\mathbf{u}/dt$ are non-zero vectors, and so we can state that $d\mathbf{u}/dt$ is perpendicular to $\mathbf{u}$.

Referring to Figure 114 again, since $\mathbf{u}$ has a constant magnitude, OP = OQ and the triangle OPQ must be isosceles. Angle QPO = $90° - \delta\theta/2$ and PQ $= 2|\mathbf{u}| \sin \delta\theta/2 = 2 \sin \delta\theta/2$; the length of PQ is the magnitude of vector $\delta\mathbf{u}$, so we can express the vector $\delta\mathbf{u}$ in terms of a unit vector $\hat{\mathbf{n}}$ along PQ thus

$$\delta\mathbf{u} = |PQ|\hat{\mathbf{n}}$$

$$= 2 \sin\frac{\delta\theta}{2}\hat{\mathbf{n}}$$

$$\frac{\delta\mathbf{u}}{\delta\theta} = \frac{\sin \delta\theta/2}{\delta\theta/2}\hat{\mathbf{n}}$$

$$\Rightarrow 1 \cdot \hat{\mathbf{n}} \quad \text{as} \quad \delta\theta \to 0$$

Since

$$\frac{\delta\mathbf{u}}{\delta t} = \frac{\delta\mathbf{u}}{\delta\theta}\frac{\delta\theta}{\delta t}$$

$$\frac{d\mathbf{u}}{dt} = \lim_{\delta t \to 0}\frac{\delta\mathbf{u}}{\delta t} = \lim_{\delta t \to 0}\frac{\delta\mathbf{u}}{\delta\theta}\frac{\delta\theta}{\delta t} = \hat{\mathbf{n}} \cdot \omega$$

$= \omega\hat{\mathbf{n}}$ and we have already shown that $d\mathbf{u}/dt$ is a vector perpendicular to $\mathbf{u}$, hence the velocity of P is always perpendicular to the radius vector $\mathbf{u}$.

The differentiation of any vector $\mathbf{a}$ results in another vector, but the differentiation of $|\mathbf{a}|$ gives another scalar. We now show the relation between them. Since

$$\frac{d}{dt}(\mathbf{a} \cdot \mathbf{b}) = \mathbf{a} \cdot \frac{d\mathbf{b}}{dt} + \mathbf{b} \cdot \frac{d\mathbf{a}}{dt} \qquad \text{(see Qu. 4, Ex. 14.)}$$

if we put $\mathbf{b} = \mathbf{a}$ then

$$\frac{d}{dt}(\mathbf{a} \cdot \mathbf{a}) = 2\mathbf{a} \cdot \frac{d\mathbf{a}}{dt}$$

But it has been shown that $\mathbf{a} \cdot \mathbf{a} = |\mathbf{a}|^2$ so that

$$\frac{d}{dt}(\mathbf{a} \cdot \mathbf{a}) = \frac{d}{dt}|\mathbf{a}|^2$$

$$= 2|\mathbf{a}|\frac{d}{dt}|\mathbf{a}| \quad \text{(Since } |\mathbf{a}| \text{ is a scalar.)}$$

Combining these two results from the differentiations

$$2|\mathbf{a}|\frac{d}{dt}|\mathbf{a}| = 2\mathbf{a} \cdot \frac{d\mathbf{a}}{dt}$$

$$\frac{d}{dt}|\mathbf{a}| = \frac{\mathbf{a}}{|\mathbf{a}|} \cdot \frac{d\mathbf{a}}{dt}$$

But $\mathbf{a}/|\mathbf{a}| = \hat{\mathbf{a}}$ (where $\hat{\mathbf{a}}$ is a unit vector in the direction of $\mathbf{a}$)

$$\frac{d}{dt}|\mathbf{a}| = \hat{\mathbf{a}} \cdot \frac{d\mathbf{a}}{dt}$$

which gives the required relation.

Vectors and mechanics
It has been established that a force is a vector quantity but the application of a force to a rigid body requires careful treatment. At once the force vector cannot be regarded as a *free* vector, its effect depends not only on magnitude and direction but also on the Point of Application.

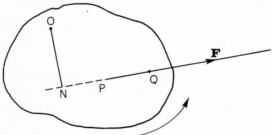

Fig. 115

The force $\mathbf{F}$ acting at point P in the rigid body pivoted at O tends to turn the body about O. This turning effect is equal in magnitude to the product of the magnitude of $\mathbf{F}$ and the perpendicular distance from O to N on the LINE OF ACTION of the force. It is clear that the same force **F** acting along the same line of action but applied at Q would have the same turning effect. When the force is localised along the same line of action its turning effect is unaltered. Forces are said to be LOCALISED VECTORS in this respect.

Again *two* forces acting at the same point in a rigid body *or* two forces whose lines of action pass through the same point, are equivalent to a single force which is the VECTOR SUM of the two forces. But if the two forces act at different points in the body then their effect is not necessarily the same as that of the force which is their vector sum.

Forces acting on a rigid body

1. Two forces acting at the same point in a rigid body have a resultant which can be found by the usual method of adding two vectors, i.e. by applying the triangle of addition. This can be verified by experiment in the laboratory.

2. Three forces acting at the same point in a rigid body have a resultant which can be found by the triangle method of addition taking the forces two at a time (or by the method of addition given earlier for vectors).

To find the resultant of a, b and c as given

Fig. 116

At O any convenient point in the plane of **a** and **b**, draw **OP** = **a**, from P draw **PQ** = **b** and from Q draw **QR** = **c**. Then from previous work the resultant is **OR**.

If the three forces are in equilibrium their resultant is zero, **OR** = **0** and R coincides with O. The three vectors form a triangle, i.e. they must be coplanar for equilibrium, and since they are all acting at the same point, they are *localised* vectors. This is known in mechanics as the triangle of forces which states that if three

141

forces acting at a point are in equilibrium they can be represented in magnitude and direction by the sides of a triangle taken in order. 3. *Polygon of forces* states that if any number of forces acting at a point can be represented in magnitude and direction by the sides of a polygon taken in order then the forces are in equilibrium.

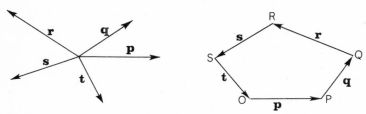

Fig. 117

Suppose the forces **p**, **q**, **r**, **s** and **t** act at the point O and are in equilibrium, then if side **OP** of the polygon represents **p** in magnitude and direction, similarly **PQ**, **QR**, **RS**, **SO** represent forces **q**, **r**, **s**, and **t**, the sides OP, PQ, RS, and SO form a closed polygon. It was shown earlier in the book that the vector sum of the sides of any closed polygon is zero hence

p + **q** + **r** + **s** + **t** = **0**

Since the sum of **p** and **q** is **OQ** and **OQ** compounded with **r** gives **OR** which in turn can be compounded with **s** to give a resultant **OT** and this force with **t** (**SO**) is zero, it follows that the forces need *not* be in the same plane.

It is important to note that the forces must act at a point and that the converse of the polygon of forces is not necessarily true.

Newton's second law

Newton's Second Law is fundamental in the study of dynamics. It states that the Rate of Change of Momentum of a body is proportional to the Impressed Force acting on that body.

Momentum is defined as the product of mass and velocity. Mass (m) is a scalar quantity but velocity (**v**) is a vector quantity so that momentum (**M**) is also a vector quantity.

M = m**v**

142

Newton's Second Law becomes

$$\mathbf{F} = \frac{d(\mathbf{M})}{dt} \quad \text{if suitable units are chosen.}$$

$$= \frac{d(m\mathbf{v})}{dt}$$

$$= m\frac{d\mathbf{v}}{dt}$$

assuming that the mass remains constant during the action,

$$= m\mathbf{a} \quad \text{since acceleration} \quad \mathbf{a} = \frac{d\mathbf{v}}{dt}$$

From previous work since vector $\mathbf{F} = m\mathbf{a}$ where m is a scalar it follows that $\mathbf{a}$ has the same direction as $\mathbf{F}$.

If the body is acted on by several forces at the same time then the acceleration produced in the body is in the same direction as the resultant of the several forces compounded vectorially.

Exercise 14a

1. Force, represented by $2\mathbf{i} + 3\mathbf{j} - 6\mathbf{k}$, acts at a point P whose position vector is $3\mathbf{i} + 4\mathbf{j} + 2\mathbf{k}$. If a second force, represented by $1\mathbf{i} + 3\mathbf{j} + 2\mathbf{k}$, also acts at this point:
 (i) what force would produce equilibrium?
 (ii) what is its line of action?

2. Forces of magnitude 3**AB** and 4**AC** act at A along the sides AB and AC of the triangle ABC. What is the magnitude and direction of the resultant?

3. In triangle ABC forces are represented by $3\overline{AB}$, $2\overline{AC}$, and $6\overline{BC}$ act along AB, AC, and BC. What is their resultant?

4. In parallelogram ABCD, the point E cuts BD in the ratio of 1:3. Forces represented by 3**AB**, 1**AD**, and 4**EC** act along the lines AB, AD, and EC. What is their resultant?

5. A body of mass 10 kg is acted upon by a force $\mathbf{F}$ represented by $2\mathbf{i} + 3\mathbf{j} + 6\mathbf{k}$, the unit of force being the Newton. What is the magnitude of the acceleration and give its direction?

6. A mass of 8 kg is acted upon by the force $x\mathbf{i} + y\mathbf{j} + z\mathbf{k}$, units being the Newton. What is the acceleration?

7. A mass m has a momentum represented by $x\mathbf{i} + y\mathbf{j} + z\mathbf{k}$, where x, y, and z are variable scalars.

   (i) What force is acting on the body?

   (ii) What is the magnitude of the acceleration of the body at time t?

   (iii) If $x = 3+2t$, $y = 1+t^2$, $z = t+t^2$, what is the acceleration at time 4 secs?

8. Forces represented by 1**AB**, 1**BC**, 2**CD**, 1**ED**, 1**EF** act along the sides of a regular hexagon. What is the resultant force?

Work done as a scalar product

If a force **F** acts on a particle which suffers a displacement **d** making an angle θ with the direction of force **F** then the work done by the force is (its component along **d**) times (the displacement).

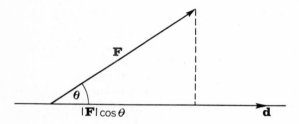

Fig. 118

Work done $= |\mathbf{F}| \cos \theta \cdot |\mathbf{d}|$

But force and displacement are vector quantities and using the DOT product:

$\Rightarrow \mathbf{F} \cdot \mathbf{d} = |\mathbf{F}||\mathbf{d}| \cos \theta = $ work done (a scalar quantity)

   So work done is the Dot (or Scalar) product of the force vector and displacement vector.

$\Rightarrow$ Work done $= \mathbf{F} \cdot \mathbf{d}$

Example 1

Find the work done when a force vector $\mathbf{F} = 4\hat{\mathbf{x}} + 3\hat{\mathbf{y}}$ acts along the displacement $\mathbf{d} = 3\hat{\mathbf{x}} + 4\hat{\mathbf{y}}$.

It has been shown that work done is the Scalar (or Dot) product of the two vectors, hence

   Work done $= \mathbf{F} \cdot \mathbf{d}$

$$= (4\hat{\mathbf{x}} + 3\hat{\mathbf{y}}) \cdot (3\hat{\mathbf{x}} + 4\hat{\mathbf{y}})$$

$$= 12 + 12 \quad \text{Since } \mathbf{x} \cdot \mathbf{y} = 0 \text{ and } \mathbf{y} \cdot \mathbf{x} = 0$$

$$= 24 \text{ units.}$$

144

Example 2

Find the work done by the force vector $\mathbf{F} = 2\mathbf{i} + 3\mathbf{j} + 7\mathbf{k}$ along the displacement vector $\mathbf{d}$ given by $\mathbf{d} = 2\mathbf{i} + 3\mathbf{j} + 5\mathbf{k}$. Since the Work Done is the Scalar quantity $= \mathbf{F} \cdot \mathbf{d}$

Work done $= (2\mathbf{i} + 3\mathbf{j} + 7\mathbf{k}) \cdot (2\mathbf{i} + 3\mathbf{j} + 5\mathbf{k})$

$= 4 + 9 + 35$ units

$= 48$ units.

Exercise 15.

1. Find the work done by a force vector represented by $\mathbf{F} = 4\mathbf{i} + 5\mathbf{j} - 2\mathbf{k}$ along the displacement vector $\mathbf{d}$ given by $\mathbf{d} = 2\mathbf{i} + 3\mathbf{j} + 4\mathbf{k}$.

2. A magnetic pole has a force acting on it represented by $\mathbf{F} = 2\mathbf{i} + 3\mathbf{j} + 5\mathbf{k}$. If the pole is moved along the displacement given by $\mathbf{d} = 1\mathbf{i} + 3\mathbf{j} + 2\mathbf{k}$ find the work done.

3. A number of constant forces $\mathbf{F}_1$, $\mathbf{F}_2$, $\mathbf{F}_3, \ldots, \mathbf{F}_n$ act on a body whose displacement is $\mathbf{d}$. Show that the work done is

$\mathbf{F}_1 \cdot \mathbf{d} + \mathbf{F}_2 \cdot \mathbf{d} + \mathbf{F}_3 \cdot \mathbf{d} + \cdots + \mathbf{F}_n \cdot \mathbf{d}$

Hence show that if $\mathbf{F}$ is the resultant force this is equivalent to $\mathbf{F} \cdot \mathbf{d}$.

4. A particle acted on by a constant force $\mathbf{P}$ moves round the sides of a square until it returns to its starting point. How much work has been done?

5. A force of 5 units acts in the direction of vector $1\mathbf{i} + 2\mathbf{j} + 3\mathbf{k}$. Give the vector representing the force. If the point of application of this force moves from the point $(0, -2, -3)$ to the point $(1, 2, 3)$, calculate the work done.

6. Three forces $\mathbf{P}_1$, $\mathbf{P}_2$, and $\mathbf{P}_3$ represented by $1\mathbf{i} + 2\mathbf{j} - \mathbf{k}$, $2\mathbf{i} - 1\mathbf{j} + \mathbf{k}$, and $2\mathbf{i} + 2\mathbf{j} + 2\mathbf{k}$ act on a body. Find a unit vector in the direction in which minimum work can be done.

7. A point P is constrained to move in the plane $2x + 2y - z + 3 = 0$. Find the unit force which acting on P produces zero work.

Turning effect of a force

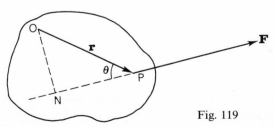

Fig. 119

Now suppose the force vector **F** acts at the point P in a body pivoted at O. The force will exert a turning moment on the body.

If **r** is the position vector of the point P then the cross (or vector) product of **r** and **F** is given by

$$\mathbf{r} \times \mathbf{F} = |\mathbf{r}||\mathbf{F}| \sin \theta \hat{\mathbf{u}}$$

where û is the unit vector forming a right-handed triple with **r** and **F**.

But

$$|\mathbf{ON}| \times |\mathbf{F}| = \text{magnitude of the turning moment}$$

$$= (|\mathbf{OP}| \sin \theta)|\mathbf{F}|$$

$$= |\mathbf{OP}||\mathbf{F}| \sin \theta = |\mathbf{r}||\mathbf{F}| \sin \theta = |\mathbf{r} \times \mathbf{F}|$$

Hence turning moment is given in magnitude by $|\mathbf{r} \times \mathbf{F}|$.

Notice that the turning moment has both magnitude and direction (i.e. clockwise or anticlockwise) and by definition must be a vector quantity. As long as the force vector **F** is localised in the line NP the moment remains constant.

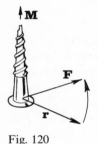

Hence using **M** to represent the turning moment as a vector, we can state that

$$\mathbf{M} = \mathbf{r} \times \mathbf{F}$$

The screw test here shows that the product of **r** on **F** gives a positive value to **M**.

Fig. 120

Angular velocity of a body

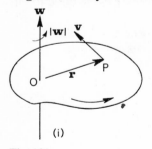

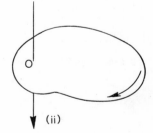

Fig. 121

If the lamina shown in the diagram is rotating anticlockwise in figure (i) about an axis through O and clockwise in figure (ii) the familiar woodscrew would tend to travel upwards in figure (i) along the axis perpendicular to the plane of motion through the centre of rotation O.

Hence in figure (ii) the rotation would be regarded as negative.

Since angular velocity has both magnitude and direction it is regarded as a vector quantity. The angular velocity has a magnitude which is defined as the rate of change of angle with time. If after time t s, the body turns through an angle of $\delta\theta$ rad in time δt s, then the average angular speed over time δt s is $\delta\theta/\delta t$. As $\delta t \to 0$ then the angular speed $|\mathbf{w}|$ at time t is

$$|\mathbf{w}| = \frac{d\theta}{dt} \quad (\text{or } \dot{\theta})$$

Consider a point P whose position vector from O is $\mathbf{r}$ and whose angular velocity measured vectorially is $\mathbf{w}$ as shown in the figure, then the velocity of the point P will be at right angles to both $\mathbf{r}$ and $\mathbf{w}$ as shown, and it can be shown that $\mathbf{v}$ is the vector product of $\mathbf{w}$ on $\mathbf{r}$, i.e.

$$\mathbf{v} = \mathbf{w} \times \mathbf{r}$$

We now show a rigid body (previously we were limited to a lamina) rotating about the axis ON

If P is a point in the rigid body let its position vector from O be $\mathbf{r}$ making an angle θ rad with the axis of rotation ON. The angular velocity $\mathbf{w}$ has magnitude $|\mathbf{w}|$ and its direction is that given by the motion of a woodscrew turning with the same rotation as $\mathbf{w}$ as shown.

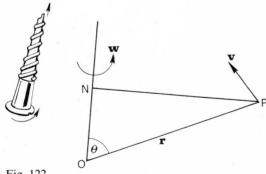

Fig. 122

The magnitude of the linear motion of P is

$$PN \times |\mathbf{w}| = OP \sin \theta |\mathbf{w}| = |\mathbf{w}||\mathbf{r}| \sin \theta = |\mathbf{w} \times \mathbf{r}|$$

6*

Hence velocity vector of P, $\mathbf{v} = \mathbf{w} \times \mathbf{r}$ in a direction at right angles to the plane containing $\mathbf{r}$ and $\mathbf{w}$ forming a right handed system.

Exercise 15a

1. A force $\mathbf{F} = 2\mathbf{i} + 3\mathbf{j} - 1\mathbf{k}$ acts through the point P(2, 3, 2). What is the moment M of the force about the point (1, 1, 1) and what is the magnitude of the moment?

2. Show that the moment $\mathbf{M}$ of a force $\mathbf{F}$ about a fixed point P is independent of the position vector $\mathbf{r}$ chosen for any point T on the line of action of $\mathbf{F}$. (*Hint:* take two points, T_1 and T_2, on the line of action of $\mathbf{F}$, let the position vectors of T_1 and T_2 be $\mathbf{r}_1$ and $\mathbf{r}_2$ show that $\mathbf{M} = (\mathbf{r}_1 \times \mathbf{F}) = (\mathbf{r}_2 \times \mathbf{F})$.)

4. A force of 5 Newtons acts in the direction A to B, where $A = (1, 1, 1)$ and $B = (3, 2, 4)$, the units being metres. What is the moment of this force about the point (0, 0, 0)?

Linear momentum

The momentum of a body is defined as the product of mass and velocity. Since mass is a scalar quantity but velocity is a vector quantity then momentum is by definition a vector quantity, call it $\mathbf{M}$

$$\mathbf{M} = m\mathbf{V}$$

The magnitude of the momentum is $|m\mathbf{v}|$ or $m|\mathbf{v}|$ and its direction and sense is that of the vector $\mathbf{v}$.

Impulse is defined as the Change of Momentum.

Hence Impulse is also a vector quantity, call it $\mathbf{I}$

$$\mathbf{I} = m\mathbf{v}_1 - m\mathbf{v}_2$$
$$= m(\mathbf{v}_1 - \mathbf{v}_2)$$

The magnitude of the Impulse $= m|\mathbf{v}_1 - \mathbf{v}_2|$, the direction is that of $(\mathbf{v}_1 - \mathbf{v}_2)$.

Since

$$\mathbf{F} = \frac{d(m\mathbf{v})}{dt} \quad \text{i.e. } \mathbf{F}\, dt = d(m\mathbf{v})$$

It can be shown that integration of vectors can be performed as with scalars, hence

$$\int_{t_1}^{t_2} \mathbf{F}\, dt = \int_{t_1}^{t_2} d(m\mathbf{v})$$

and assuming that the force **F** is constant during the time $(t_2 - t_1)$

$$\mathbf{F}(t_2 - t_1) = \left[m\mathbf{v} \right]_{t_1}^{t_2} = m\mathbf{v}_2 - m\mathbf{v}_1$$

Putting $t = (t_2 - t_1)$ gives

$$\mathbf{F}t = m\mathbf{v}_2 - m\mathbf{v}_1$$

But Impulse is defined as the change of momentum and so

Impulse $= \mathbf{F}t$

Since the time t is a scalar quantity and force **F** is a vector quantity then this relation shows that the vector quantity impulse has the same direction and sense as the vector force producing it.

Angular Momentum or moment of momentum
If a body is rotating about an axis and at any moment a particle of mass m, position vector **r**, has a velocity $\dot{\mathbf{r}}$, then the momentum of the particle is $m\dot{\mathbf{r}}$ and the moment of the momentum $m\dot{\mathbf{r}}$ about the axis is given by $\mathbf{r}\times(m\dot{\mathbf{r}})$. The moment of momentum of the whole body is $\sum (\mathbf{r}\times m\dot{\mathbf{r}})$. The angular Momentum or moment of momentum is usually denoted by **H**. That it is a vector quantity can be demonstrated in the laboratory by filling the tube of a bicycle wheel with lead and then showing that the wheel when set into motion can be used to show that angular momentum has both magnitude and direction. The experiments are described in a number of standard books on physics. It can also be shown by experiment that any change in the magnitude or direction of the angular momentum causes a couple or torque to be set up.

Consider a lamina of the body perpendicular to the axis of rotation and rotating with constant angular velocity ω. The velocity of any particle of position vector **r** will be $\dot{\mathbf{r}}$ and perpendicular to the direction of **r**. Also

$$|\dot{\mathbf{r}}| = |\mathbf{r}|\omega$$

therefore the angular momentum of the particle will be $(\mathbf{r}\times m\dot{\mathbf{r}})$ and

$$\begin{aligned}(\mathbf{r}\times m\dot{\mathbf{r}}) &= |\mathbf{r}||m\dot{\mathbf{r}}|\hat{\mathbf{n}} \\ &= |\mathbf{r}|\, m\, |\dot{\mathbf{r}}|\hat{\mathbf{n}} \\ &= |\mathbf{r}|\, m\, |\mathbf{r}|\omega\hat{\mathbf{n}}\end{aligned}$$

where $\hat{\mathbf{n}}$ is the unit vector perpendicular to both **r** and $\dot{\mathbf{r}}$, i.e., along the axis of rotation.

It follows that the magnitude of the angular momentum of the whole lamina is

$$\sum m|\mathbf{r}|^2 \times \omega = I\omega$$

(where I is the Moment of Inertia so for the whole body $= I\dot{\theta}$ of the lamina.)

Motion in a circle

The motion of a particle in a circle of radius a has two important properties:
1. The radius vector $\mathbf{r}$ remains of constant magnitude, if the origin is taken at the centre.
2. The motion is in one plane.
The Scalar product of the radius vector with itself

$$\mathbf{r} \cdot \mathbf{r} = |\mathbf{r}|^2 = a^2 \tag{i}$$

and the scalar product of $\mathbf{r}$ with a unit vector $\hat{\mathbf{n}}$ normal to the plane of the circle must be zero since $\mathbf{r}$ and $\hat{\mathbf{n}}$ must always be orthogonal.

$$\mathbf{r} \cdot \hat{\mathbf{n}} = 0 \tag{ii}$$

Differentiation of equation (i) with respect to time gives

$$\frac{d}{dt}(\mathbf{r} \cdot \mathbf{r}) = \mathbf{r} \cdot \frac{d}{dt}(\mathbf{r}) + \frac{d}{dt}(\mathbf{r}) \cdot \mathbf{r}$$

$$= 2\mathbf{r} \cdot \dot{\mathbf{r}}$$

$$= \frac{d}{dt}(a^2)$$

$$= 0$$

This means that vectors $\mathbf{r}$ and $\dot{\mathbf{r}}$ are perpendicular to each other.
Differentiation of equation (ii) with respect to time gives

$$\frac{d}{dt}(\mathbf{r} \cdot \hat{\mathbf{n}}) = 0$$

$$\frac{d}{dt}(\mathbf{r}) \cdot \hat{\mathbf{n}} + \mathbf{r} \cdot \frac{d}{dt}(\hat{\mathbf{n}}) = 0$$

150

but since $\hat{\mathbf{n}}$ is of unit magnitude and constant direction $(d/dt)(\hat{\mathbf{n}}) = 0$

hence $\hat{\mathbf{n}} \cdot \dot{\mathbf{r}} = 0$

Thus $\dot{\mathbf{r}}$ is perpendicular also to $\hat{\mathbf{n}}$ as well as to $\mathbf{r}$, hence $\dot{\mathbf{r}}$ lies in the plane of $\mathbf{r}$ and is perpendicular to $\mathbf{r}$. $\mathbf{r}$, $\hat{\mathbf{n}}$ and $\dot{\mathbf{r}}$ are mutually orthogonal.

Next consider the vector product $(\hat{\mathbf{n}} \times \mathbf{r})$

$$\hat{\mathbf{n}} \times \mathbf{r} = |\hat{\mathbf{n}}||\mathbf{r}| \sin \frac{\pi}{2} \hat{\mathbf{s}}$$

where $\hat{\mathbf{s}}$ is a unit vector orthogonal to $\hat{\mathbf{n}}$ and $\mathbf{r}$

$$= |\mathbf{r}|\hat{\mathbf{s}} \quad \text{since} \quad |\hat{\mathbf{n}}| = 1 \quad \text{and} \quad \sin \frac{\pi}{2} = 1$$

The vector $|\mathbf{r}|\hat{\mathbf{s}}$ is orthogonal to $\hat{\mathbf{n}}$ and $\mathbf{r}$ and so is parallel to $\dot{\mathbf{r}}$. We can state

$$\dot{\mathbf{r}} = \omega|\mathbf{r}|\hat{\mathbf{s}}$$

where ω is a scalar which need not be constant.

It was shown earlier that the

speed $|\dot{\mathbf{r}}|$ = Angular speed times the radius

so that the scalar ω is the angular velocity of the particle and $\omega = \dot{\theta}$ where θ is the angle in radians measured from some fixed base direction.

It follows that

$\dot{\mathbf{r}} = \omega\hat{\mathbf{n}} \times \mathbf{r}$ and this is illustrated in Fig. 123

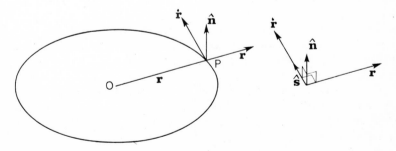

Fig. 123

The vector product of $\hat{\mathbf{n}}$ on $\mathbf{r}$ produces the vector $\omega|\mathbf{r}|\hat{\mathbf{s}}$ in the direction shown, the order is important, the vector product of $\mathbf{r}$

on $\hat{\mathbf{n}}$ would be in the opposite direction, i.e. $-\dot{\mathbf{r}}$. $\hat{\mathbf{n}}$, $\mathbf{r}$ and $\dot{\mathbf{r}}$ form a right-handed system.

The order in vector products must be preserved and this is important in the differentiation of the vector product.

$$\dot{\mathbf{r}} = \omega\hat{\mathbf{n}} \times \mathbf{r}$$

Hence

$$\frac{d}{dt}(\dot{\mathbf{r}}) = \frac{d}{dt}(\omega\hat{\mathbf{n}}) \times \mathbf{r} + \omega\hat{\mathbf{n}} \times \frac{d}{dt}(\mathbf{r})$$

i.e.

$$\ddot{\mathbf{r}} = \dot{\omega}\hat{\mathbf{n}} \times \mathbf{r} + \omega\hat{\mathbf{n}} \times \dot{\mathbf{r}} \quad \text{But} \quad \dot{\mathbf{r}} = \omega\hat{\mathbf{n}} \times \mathbf{r}$$

$$= \dot{\omega}\hat{\mathbf{n}} \times \mathbf{r} + \omega\hat{\mathbf{n}} \times (\omega\hat{\mathbf{n}} \times \mathbf{r})$$

$$= \dot{\omega}\hat{\mathbf{n}} \times \mathbf{r} + \omega^2[\hat{\mathbf{n}} \times (\hat{\mathbf{n}} \times \mathbf{r})]$$

$$= \dot{\omega}\hat{\mathbf{n}} \times \mathbf{r} + \omega^2[(\hat{\mathbf{n}} \cdot \mathbf{r})\hat{\mathbf{n}} - |\hat{\mathbf{n}}|^2\mathbf{r}] \quad \text{and} \quad \hat{\mathbf{n}} \cdot \mathbf{r} = 0, \quad |\hat{\mathbf{n}}|^2 = 1$$

$$\text{since } \hat{\mathbf{n}} \cdot \hat{\mathbf{n}} = |\hat{\mathbf{n}}|^2$$

$$= \dot{\omega}\hat{\mathbf{n}} \times \mathbf{r} - \omega^2\mathbf{r}$$

The acceleration $\ddot{\mathbf{r}}$ has two components. The first, $\dot{\omega}|\hat{\mathbf{n}} \times \mathbf{r}|$ or $\dot{\omega}|\mathbf{r}|$, is in the direction of the velocity $\dot{\mathbf{r}}$ i.e. in the plane of the circle but orthogonal to $\hat{\mathbf{n}}$ and $\mathbf{r}$, called the TRANSVERSE direction. Since $|\mathbf{r}| = a$ the transverse component of acceleration has magnitude $\dot{\omega}a$.

The second component $-\omega^2|\mathbf{r}|$ has the direction of $-\mathbf{r}$ i.e. it is directed along the radius towards the centre of the circle and is called the RADIAL component of the acceleration.

In the case of motion in a circle the transverse direction is tangential to the circle and the radial direction is normal to the circle so the tangential and normal components of acceleration are also of magnitude $\dot{\omega}a$ and ω^2a.

If motion along a *plane curve* is considered it is clear that the transverse direction which is orthogonal to the position vector and lies in the plane is not in general the tangent to the curve as in the case of motion in a circle. Similarly the radial direction is not generally the normal direction to the curve.

152

Tangential and normal components of velocity and acceleration

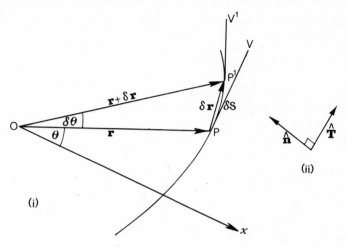

Fig. 124

P is a particle moving along the space curve shown, O is any convenient origin and **r** is the position vector of P. If PV is the geometrical tangent to the curve at P it is clear that when P moves to the new position P^1 that the tangent P^1V^1 to the curve at P^1 has a new direction which varies with the curvature of the length of the arc PP^1.

Let $\hat{\mathbf{T}}$ be the unit vector along the tangent at P. Since $\hat{\mathbf{T}}$ has unit magnitude although its direction varies with s, (where s is the length of the curve measured from some fixed point) then

$$\hat{\mathbf{T}}.\hat{\mathbf{T}} = 1 \quad \text{since} \quad \hat{\mathbf{T}}.\hat{\mathbf{T}} = |\hat{\mathbf{T}}|^2 = 1$$

Differentiating with respect to s

$$\frac{d\hat{\mathbf{T}}}{ds}.\hat{\mathbf{T}} + \hat{\mathbf{T}}.\frac{d\hat{\mathbf{T}}}{ds} = 0$$

$$2\hat{\mathbf{T}}.\frac{d\hat{\mathbf{T}}}{ds} = 0$$

hence $d\hat{\mathbf{T}}/ds$ is a vector perpendicular to $\hat{\mathbf{T}}$ and making an angle of $\pi/2$ in the positive direction from $\hat{\mathbf{T}}$ as shówn in Fig. (ii). Let $\hat{\mathbf{n}}$ be

153

the unit vector in this direction then

$$\frac{d\hat{\mathbf{T}}}{ds} = K\hat{\mathbf{n}} \quad \text{where K is defined as the curvature and}$$

$$\rho = \frac{1}{K} \quad \text{is the radius of curvature}$$

$\hat{\mathbf{n}}$ is the normal to the curve at P.
From the definition of the differentiation of a vector

$$\frac{d\mathbf{r}}{dt} = \lim_{\delta t \to 0} \frac{(\mathbf{r} + \delta\mathbf{r}) - \mathbf{r}}{\delta t} = \lim_{\delta t \to 0} \frac{\delta\mathbf{r}}{\delta t}$$

as $\delta t \to 0$ the magnitude of vector $\mathbf{PP}^1(\delta\mathbf{r})$ tends to become equal to the arc length δs and have the direction of the unit tangent vector $\hat{\mathbf{T}}$,

$$\Rightarrow \lim_{\delta t \to 0} \frac{\delta\mathbf{r}}{\delta t} \to \lim_{\delta t \to 0} \frac{\delta s}{\delta t}\hat{\mathbf{T}} = |\mathbf{v}|\hat{\mathbf{T}}$$

where $|\mathbf{v}|$ is the magnitude of the velocity of the point P.
But $d\mathbf{r}/dt$ is the velocity of the point P

$$\Rightarrow \mathbf{v} = |\mathbf{v}|\hat{\mathbf{T}}$$

By differentiation of $\mathbf{v}$ with respect to time we can find the acceleration,

$$\mathbf{a} = \frac{d\mathbf{v}}{dt} = \frac{d}{dt}(|\mathbf{v}|\hat{\mathbf{T}}) = \frac{d(|\mathbf{v}|)}{dt}\hat{\mathbf{T}} + |\mathbf{v}|\frac{d\hat{\mathbf{T}}}{dt}$$

$$= \frac{d|\mathbf{v}|}{dt}\hat{\mathbf{T}} + |\mathbf{v}|\frac{d\hat{\mathbf{T}}}{ds}\frac{ds}{dt}$$

$$= \frac{d|\mathbf{v}|}{dt}\hat{\mathbf{T}} + |\mathbf{v}|^2\frac{d\hat{\mathbf{T}}}{ds}$$

$$= \frac{d|\mathbf{v}|}{dt}\hat{\mathbf{T}} + |\mathbf{v}|^2 K\hat{\mathbf{n}}$$

$$= \frac{d|\mathbf{v}|}{dt}\hat{\mathbf{T}} + \frac{|\mathbf{v}|^2}{\rho}\hat{\mathbf{n}}$$

It is seen that the acceleration has two components (i) $d(|\mathbf{v}|)/dt$ along the tangent to the curve at P (ii) $|\mathbf{v}|^2/\rho$ along the normal to the curve.

Central forces and orbits

If a particle moves in space when acted upon by a force directed towards some fixed point, the path of the particle is called its ORBIT. The force is called the CENTRAL FORCE and the fixed point is the CENTRE OF FORCE.

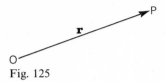

Fig. 125

Suppose the particle at P is of mass m and is acted upon by a single force directed towards the fixed point O which is taken as origin. Then **OP** is the position vector of P and is represented by **r**. Since the single force acting is a vector in the same direction as the vector **r** we know that it can be represented by $\lambda\mathbf{r}$ where λ is a scalar not necessarily constant which can be positive or negative (in this case it is negative since the force acts from P towards O whereas **r** has the direction from O to P). From Newton's Second Law of Motion we know that Mass × Acceleration = Force, hence

$$m\ddot{\mathbf{r}} = \lambda\mathbf{r} \quad \left(\text{where } \ddot{\mathbf{r}} \text{ means } \frac{d^2\mathbf{r}}{dt^2}\right)$$

The vector product with **r** gives

$$m\mathbf{r}\times\ddot{\mathbf{r}} = \lambda\mathbf{r}\times\mathbf{r} = 0$$

since $\lambda\mathbf{r}$ and **r** have the same direction.
Next consider the differentiation of $m\mathbf{r}\times\dot{\mathbf{r}}$ with respect to t

$$\frac{d}{dt}(m\mathbf{r}\times\dot{\mathbf{r}}) = m\mathbf{r}\times\frac{d}{dt}(\dot{\mathbf{r}}) + m\dot{\mathbf{r}}\times\frac{d\mathbf{r}}{dt}$$

but $m\dot{\mathbf{r}}$ and $\dot{\mathbf{r}}$ have the same direction hence $m\dot{\mathbf{r}}\times\dot{\mathbf{r}} = 0$ and $m\mathbf{r}\times\ddot{\mathbf{r}}$ has been shown equal to **0** also, hence

$$\frac{d}{dt}(m\mathbf{r}\times\dot{\mathbf{r}}) = 0$$

Integration with respect to t gives

$$\int\frac{d}{dt}(m\mathbf{r}\times\dot{\mathbf{r}})\,dt = \int\mathbf{0}\,dt = \mathbf{k}$$

where **k** is a constant vector.

$$\Rightarrow m\mathbf{r}\times\dot{\mathbf{r}} = \mathbf{k}$$

This means that the constant vector **k** is at right angles to the plane containing $m\mathbf{r}$ and $\dot{\mathbf{r}}$.

155

Hence the position vector **r** is always normal to the constant vector **k** and the point P must always lie in a plane since its position vector is bound to the centre of force and is always normal to the constant vector. The path of P is a plane curve which is determined by the initial velocity and position of the particle.

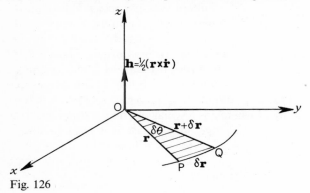

Fig. 126

Figure 126 shows a particle P moving along a plane curve under the action of a central force. In the interval of time δt the particle moves from P to Q as shown. The area of segment OPQ is approximately $\frac{1}{2}$OP . PQ sin $\hat{OPQ}$. Calling δA the area of segment OPQ

$$\delta A \simeq \tfrac{1}{2}|\mathbf{r}||\delta\mathbf{r}| \sin \hat{OPQ}$$

dividing by δt

$$\frac{\delta A}{\delta t} \simeq \tfrac{1}{2}|\mathbf{r}|\left|\frac{\delta\mathbf{r}}{\delta t}\right| \sin \hat{OPQ}$$

as $\delta t \to 0$

$$\lim_{\delta t \to 0} \frac{\delta A}{\delta t} = \frac{dA}{dt} = \lim_{\delta t \to 0} \tfrac{1}{2}|\mathbf{r}| \frac{\delta\mathbf{r}}{\delta t} \sin \hat{OPQ}$$
$$= \tfrac{1}{2}|(\mathbf{r} \times \dot{\mathbf{r}})|$$

But we have shown earlier that when a particle moves under a central force that $m\mathbf{r} \times \dot{\mathbf{r}}$ is equal to a constant vector, hence $\mathbf{r} \times \dot{\mathbf{r}}$ has a constant magnitude and this means that the rate of change of area swept out by the position vector is constant.

In 1609 Kepler deduced this result empirically for the motion of the planets round the sun. He stated that the motion of the planets resulted in equal areas being swept out in equal times by the imaginary line drawn from the planet to the sun. He also stated that the orbits of the planets are ellipses with the sun as one of the foci. With the help of Kepler's work Newton was able to formulate his theory of universal gravitation that two masses m and M are attracted to each other by a force **F** such that

$$|\mathbf{F}| = \frac{kMm}{r^2}$$

where k is a universal constant and r is the distance between their centres. Assuming that the influence of the other planets can be neglected and using Newton's Law it can be proved that if the sun is taken as origin then the path of a planet round the sun is an elliptical orbit with the sun as the focus. The proof can be found in more advanced books on gravitational theory.

The study of space curves and surfaces has had renewed interest in recent years as space travel and space research have become more widely understood. The elliptical orbits of the sputniks and satellites have often been illustrated in newspapers and magazines but it would be useful here to point out that the plane in which these objects orbit round the earth is determined beforehand by their speed and direction at launching.

The space curves of capsules launched towards the moon and other planets are complicated examples of differential geometry.

Exercise 16.

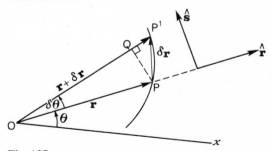

Fig. 127

The point P moves along the plane curve as shown. The origin O is chosen at any point in the plane of the curve and the position vector of P is **r**, the angle θ is measured from the fixed direction

157

An Introduction to Vectors

of Ox. The unit vector in the direction of **OP** is $\hat{\mathbf{r}}$ and the unit vector perpendicular to **OP** but in the plane of the curve is $\hat{\mathbf{s}}$.

1. Show that $d\hat{\mathbf{r}}/dt = \dot{\theta}\hat{\mathbf{s}}$ or $\omega\hat{\mathbf{s}}$.

2. Expressing $\mathbf{r} = |\mathbf{r}|\hat{\mathbf{r}}$ show that the radial and transverse components of velocity are $d|\mathbf{r}|/dt$ and $|\mathbf{r}|\dot{\theta}$ or $|\mathbf{r}|\omega$.

3. Using the result of Qu. 2 show that the radial and transverse components of acceleration are

$$\frac{d}{dt}|\dot{\mathbf{r}}| - |\mathbf{r}|\dot{\theta}^2 \quad \text{and} \quad 2|\dot{\mathbf{r}}|\dot{\theta} + |\mathbf{r}|\ddot{\theta}$$

4. If the curve has the equation $|\mathbf{r}| = a \sin kt$ where a and k are constants, then $\mathbf{r} = a \sin kt\hat{\mathbf{r}}$. Find the radial and transverse components of velocity at time $t = 1/k$.

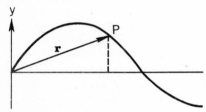

Fig. 128

5. If the curve referred to coordinate axes has the parametric equations $y = a \sin kt$, $x = kt$ then $\mathbf{r}$ the position vector is given by $\mathbf{r} = x\mathbf{i} + y\mathbf{j}$. Find the velocity vector and acceleration vector of P at time t.

6. A particle P is moving along a curve so that its position vector $\mathbf{r}$ is given by $\mathbf{r} = a \cos \omega t\mathbf{i} + a \sin \omega t\mathbf{j}$.

 (i) Find the velocity at time t and give its magnitude.

 (ii) Find the unit tangent vector to the curve at time t.

 (iii) Find the unit vector normal to the curve at time t.

 (iv) Find the acceleration at time t and give its tangential and normal components.

7. A particle P moves along a curve such that at time t its position (x, y) is given by $x = akt$, $y = a \sin kt$.

 (i) Give the position vector of P.

 (ii) Find the velocity and acceleration of P.

 (iii) Find the component of acceleration along the tangent to the curve at P.

8. A point P moves round the cardioid so that its position vector

158

$\mathbf{r} = a(\cos\theta + 1)\hat{\mathbf{r}}$ where $\dot{\theta} = \omega = $ constant. Find the radial and transverse components of acceleration.

9. A particle of mass m and moving with velocity $\mathbf{v}$ and acted upon by a force $\mathbf{F}$, has a position vector $\mathbf{r}$ relative to a convenient origin O. The moment of $\mathbf{F}$ about O i.e. the torque is given by $\mathbf{M} = \mathbf{r} \times \mathbf{F}$ and the moment of momentum (angular momentum) $\mathbf{H}$ is given by $\mathbf{H} = \mathbf{r} \times m\mathbf{v}$. Show that $\mathbf{M} = d\mathbf{H}/dt$.

10. A particle P moving with velocity $\mathbf{v}$ is attracted to a fixed point O by a single force $f = (k/\mathrm{OP}^2)\mathbf{r}$. Show that $\mathbf{r} \times \mathbf{v} = \mathbf{h}$ where $\mathbf{h}$ is a constant vector, hence prove that the angular momentum is constant.

11. A particle moves so that its position vector $\mathbf{r}$ is given by $\mathbf{r} = k\cos\theta\mathbf{a} + m\sin\theta\mathbf{b}$. Identify the motion if $\mathbf{a}$ and $\mathbf{b}$ are constant vectors and $\theta = pt$ where p is also constant.

12. A body of mass M is constrained to move in the direction given by vector $\mathbf{d} = 6\mathbf{i} + 3\mathbf{j} + 2\mathbf{k}$. A force represented by $\mathbf{F} = 2\mathbf{i} + 4\mathbf{j} + 1\mathbf{k}$ acts on the body for 4 sec. If the initial velocity of the body is 5 units/sec in the direction of $\mathbf{d}$ what is its final velocity?

13. At a certain moment a body of mass 10 kg has a velocity vector $\mathbf{v}_1 = 1\mathbf{i} + 2\mathbf{j} + 3\mathbf{k}$ the unit of length being metres, time in seconds. It is acted upon by a force represented by the vector $\mathbf{F}_1 = 2\mathbf{i} + 3\mathbf{j} + 1\mathbf{k}$ (the unit being the Newton) and a second force represented by $\mathbf{F}_2 = 1\mathbf{i} + 2\mathbf{j} + 3\mathbf{k}$. What is the velocity vector after 10 secs? What are the initial and final speeds?

Chapter 11
Vector algebra as an abstract algebra

So far the approach to vectors has been practical, through numerical problems, practical applications to plane geometry, analytical geometry and mechanics. The historical development of the work that led to the extension of the concept of multiplication in number algebra, to cover scalar product and vector product of two vectors, was intended to explain how these processes had their origins in practical situations.

In an abstract branch of mathematics, axioms or laws are stated and the subject developed from this basis without the guidance of practical applications. The laws of vector algebra given on page 25 for 2-vectors were shown to be valid for 3-vectors. If the laws stated are accepted for vectors of n-dimensional space then immediately we are working in an abstract algebra. The mathematician can think of the magnitude of an n-dimensional vector although he cannot make a geometrical or *real* model to show it.

The equation $x^2 + y^2 = a^2$ represents a circle in two-dimensional space and $x^2 + y^2 + z^2 = a^2$ represents a sphere in three-dimensional space, but we can state that if $x_1^2 + x_2^2 + x_3^2 + x_4^2 \ldots x_n^2 = a^2$ this represents a 'hypersphere' in n-dimensional space. Vectors of four or more dimensions can be added, and the process will be defined as on page 80.

Suppose $\mathbf{a} = \begin{bmatrix} a_1 \\ a_2 \\ a_3 \\ \vdots \\ a_n \end{bmatrix}$ and $\mathbf{b} = \begin{bmatrix} b_1 \\ b_2 \\ b_3 \\ \vdots \\ b_n \end{bmatrix}$

where $(a_1, a_2, a_3, \ldots a_n)$ and $(b_1, b_2, b_3, \ldots b_n)$ are the components of the vectors. It is impossible to imagine any process of addition which can be set out geometrically like the triangle of vectors—a geometrical model of the situation cannot be devised but the algebraic process of adding components is very suitable.

Then the sum of **a** and **b** is defined as

$$\mathbf{a+b} = \begin{bmatrix} a_1+b_1 \\ a_2+b_2 \\ a_3+b_3 \\ \vdots \\ a_n+b_n \end{bmatrix}$$

Scalar multiplication of **a** by m will be defined

$$m\mathbf{a} = \begin{bmatrix} ma_1 \\ ma_2 \\ ma_3 \\ \vdots \\ ma_n \end{bmatrix} = m \begin{bmatrix} a_1 \\ a_2 \\ a_3 \\ \vdots \\ a_n \end{bmatrix}$$

$$|\mathbf{a}| = \sqrt{a_1^2 + a_2^2 + a_3^2 \ldots a_n^2}$$

where $|\mathbf{a}|$ is defined as the modulus.

It is often more convenient to express a vector algebraically as a row vector, rather than in the column form. Thus $\mathbf{a} = (a_1, a_2, a_3 \ldots a_n)$ and $\mathbf{b} = (b_1, b_2, b_3 \ldots b_n)$.

Generally a vector **V** is expressed

$$\mathbf{V} = (x_1, x_2, x_3, \ldots x_n)$$

In the early parts of this book vectors were represented by geometrical two-dimensional models, that is by directed line segments. These are often called 2-vectors and we then showed that by imposing an orthogonal cartesian basis on the plane of the vectors they could be represented algebraically using first their components in the base directions and secondly by a pair of numbers only, derived from the components. Throughout the following treatment, the geometric model or the algebraic model was used, whichever was the more convenient. These ideas were then extended to a space of three dimensions and again it was found convenient to use either the geometrical model or the algebraic model to represent the vector and increasingly it was often found useful for the treatment to be a mixture of the two ideas as on pages 106–107.

Directed line-segments give a geometrical model of a vector but column vectors or row vectors give a numerical or algebraic model of a vector. In the abstract, vectors are often regarded as 'mathematical quantities' which require two or more numbers to define them. The numbers or elements can be selected from the real number field and if the scalars we use for multiplication are drawn from the same field then we are dealing with *real vectors* in *real space*. (If the numbers are drawn from the field of complex numbers then the vectors are over complex space and outside the scope of this introduction.)

The base orthogonal unit vectors **i**, **j** and **k** of three dimensional space have coordinates $(1, 0, 0)$, $(0, 1, 0)$ and $(0, 0, 1)$. Extending this idea to four-dimensional space gives the corresponding base vectors to span the space as $(1, 0, 0, 0)$, $(0, 1, 0, 0)$, $(0, 0, 1, 0)$ and $(0, 0, 0, 1)$. The idea can be extended to a space of n-dimensions for which no geometrical models can be constructed.

Linear algebra
Consider the linear equation

$$2x - 3y + 3z = 0$$

Obviously $x = 3$, $y = 4$, $z = 2$ satisfies this equation.

The vector $(3, 4, 2)$ is called a *solution vector* of the system $2x - 3y + 3z = 0$.

But obviously $(3k, 4k, 2k)$ or $k(3, 4, 2)$ is also a solution vector of the same system and is a scalar multiple of the first solution vector $(3, 4, 2)$. The linear equation $2x - 3y + 3z = 0$ represents a plane through the origin and every solution vector lies in this plane. We say that the solution space of the equation $2x - 3y + 3z = 0$ is the plane through the origin which contains the vector $(3, 4, 2)$. The solution $x = 0$, $y = 0$, $z = 0$ is called a *trivial* solution of the equation and is usually of little practical value.

Next we see that the vector $(3, 3, 1)$ is also a solution vector of the equation.

Calling vector $(3, 4, 2)$ the solution vector **S** and vector $(3, 3, 1)$ the solution vector **T** we now show that the vector $\mathbf{V} = \mathbf{S} + \mathbf{T}$ is also a solution vector.

$$\mathbf{V} = \mathbf{S} + \mathbf{T} = (3+3, 4+3, 2+1) = (6, 7, 3)$$

and

$$2(6) - 3(7) + 3(3) = 0$$

162

hence **V** is a solution vector. Every solution vector of $2x - 3y + 4z = 0$ could be expressed as a linear combination of S and T. For example $(-3, +3, +5)$ is another solution vector and let

$(-3, +3, +5) = aS + bT$

$$= a(3, 4, 2) + b(3, 3, 1)$$

$$= (3a + 3b), (4a + 3b), (2a + b)$$

$\Rightarrow 3a + 3b = -3, 4a + 3b = +3, 2a + b = +5$

$\Rightarrow a = 6, b = -7$

The solution vector $(-3, +3, +5)$ is $6(3, 4, 2) - 7(3, 3, 1)$, a linear combination of $(3, 4, 2)$ and $(3, 3, 1)$.

These ideas can be extended to linear equations of any number of variables usually referred to as linear algebra dealing with n-dimensional vector space.

From previous work we know that the equation we have been considering $2x - 3y + 3z = 0$ which passes through the origin, has a vector $\mathbf{u} = 2\mathbf{i} - 3\mathbf{j} + 3\mathbf{k}$ normal to it. If we now consider also the equation $x - 2y + 3z = 3$, we know that this equation represents another plane. The vector normal to this plane is $\mathbf{v} = 1\mathbf{i} - 2\mathbf{j} + 3\mathbf{k}$. Since vectors **u** and **v** are not parallel then the planes to which they are normal cannot be parallel. The planes therefore intersect in a straight line in three dimensions. The solution vector which satisfies both the equations

$2x - 3y + 3z = 0$

$x - 2y + 3z = 3$

must lie along the straight line in which they intersect and this means that x, y and z can only be expressed in terms of some parameter t. The solution vector $(-6, -3, 1)$ satisfies both equations and every vector of the form $(t, t + 3, (t + 9)/3)$ will satisfy the two equations because all these vectors lie along the straight line in which the planes intersect.

Next consider the equation $-2x + y - 4z = 19$. The vector normal to the plane it represents is given by $\mathbf{w} = -2\mathbf{i} + 1\mathbf{j} - 4\mathbf{k}$. Since **w** is not parallel to either **u** or **v** the three planes are non-parallel, each pair of planes will intersect in a straight line, there will be three such straight lines and it is obvious that unless these three straight lines are concurrent there will be no unique solution.

In fact the three straight lines are concurrent at the point $(-12, -9, -1)$.

When dealing with linear equations with two variables or with three variables we can construct geometrical models to illustrate the problems. When the linear equations have four or more variables then we can no longer construct such models and methods using matrices then become of great value, making use of abstract vector algebra.

Vector spaces

The purely algebraic structure which we extract from the study of vectors treated as line segments (i.e. vector analysis) forms the basis of the extension to more generalised abstract vectors which we term vector spaces. The system of complex numbers can be taken as an illustration of the mathematical structure of a vector space.

It has been shown earlier that the sum of two free vectors in a plane is another vector in the same plane. Similarly the sum of two free vectors in space (three dimensions) is another vector in the same space. We say that such a system is *closed* under the process of addition. Further we showed that the process of addition is *commutative* and *associative*. We further showed that such vectors can be multiplied by a scalar number to produce another vector in the same system. We can extend these ideas to abstract mathematical quantities which are also called VECTORS, where we can take ordered pairs or ordered triples of numbers for example, as the ELEMENTS of this vector system which obey the laws given for closure, commutativity, associativity and scalar multiplication, and then such a system is called a VECTOR SPACE.

Using bold capital letters to represent vectors and small letters to represent scalars we can summarise the structure as follows:

I $\mathbf{A} + \mathbf{B} = \mathbf{C}$ (Closure)

II $\mathbf{A} + \mathbf{B} = \mathbf{B} + \mathbf{A}$ (Commutative)

III $\mathbf{A} + (\mathbf{B} + \mathbf{C}) = (\mathbf{A} + \mathbf{B}) + \mathbf{C}$ (Associative)

IV $\mathbf{A} + (-\mathbf{A}) = \mathbf{0}$ (Inverse for addition)

V $\mathbf{A} + \mathbf{0} = \mathbf{A}$

Scalar multiplication follows these laws:

VI $s\mathbf{A}$ is a uniquely defined element.

VII $s(\mathbf{A}+\mathbf{B}) = s\mathbf{A}+s\mathbf{B}$ ⎫
VIII $(s+t)\mathbf{A} = s\mathbf{A}+t\mathbf{A}$ ⎬ (Distributive Laws)
⎭

  IX $s(t\mathbf{A}) = (st)\mathbf{A}$

  X $1 \cdot \mathbf{A} = \mathbf{A}$ (Unity element for scalar multiplication)

Having summarised the mathematical structure of a vector space we see that the real vectors such as displacements, velocities, forces, etc., of the early part of this book satisfy the conditions for a vector space.

We can define complex numbers in two ways:

1. If x and y are real numbers and $i^2 = -1$ (or $i = \sqrt{-1}$) then $x+iy$ is called a complex number. x is called the *real* part of the complex number and iy the *imaginary* part, the abstract 'number' i is defined to follow the ordinary arithmetical processes.

$$x+i = i+x$$

where x is a real number

$$xi = ix$$

The sum of two complex numbers $a+ib$ and $c+id$ is given by

$$(a+ib)+(c+id) = (a+c)+i(b+d)$$

i.e. the sum of two complex numbers is another complex number. The difference of two complex numbers is

$$(a+ib)-(c+id) = (a-c)+i(b-d)$$

also a complex number. Notice that

$$(a+ib)+(c+id) = (a+c)+i(b+d)$$
$$= (c+a)+i(d+b)$$
$$= (c+id)+(a+ib)$$

and addition must therefore be *commutative*. Again

$$(a+ib)+\{(c+id)+(e+if)\}$$
$$= (a+ib)+(c+e)+i(f+d)$$

165

$$= (a+c+e)+i(b+f+d)$$
$$= (a+c)+i(b+d)+(e+if)$$
$$= \{(a+ib)+(c+id)\}+(e+if)$$

which means that addition is *associative*. If the 'number' i is defined to follow normal arithmetical processes then

$$(a+ib)(c+id) = a(c+id)+ib(c+id)$$
$$= ac+iad+ibc+i^2bd$$
$$= ac+iad+ibc+(-1)bd$$
$$= (ac-bd)+i(ad+bc)$$

If s is a real number then

$$s(a+ib) = sa+isb$$

and if t is another real number then

$$(s+t)(a+ib) = (s+t)a+(s+t)ib$$
$$= sa+ta+sib+tib$$
$$= s(a+ib)+t(a+ib)$$

also

$$s[t(a+ib)] = s[at+ibt]$$
$$= ast+ibst$$
$$= t(as+ibs)$$
$$= t[s(a+ib)]$$

The complex number $x+iy$ is not the same as $y+ix$, x and y are real numbers so that the *order* of the two real numbers (x, y) determines the complex number which they define.

In the x, y plane the point P defined by x and y as *coordinates in that order*, is unique. It is taken to represent the complex number $x+iy$ since Argand showed that iy can be taken to mean a direction taken at right angles to the x-axis in an anti-clockwise direction.

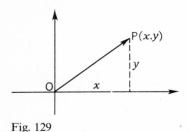

The position vector of P is **OP** so that *associated* with each point P in the plane is a unique position vector of components x and y. Hence we can associate a vector in the x, y plane with each complex number.

Fig. 129

The addition of two complex numbers $a+ib$ and $c+id$

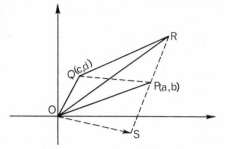

Fig. 130

By definition

$$(a+ib)+(c+id) = (a+c)+i(b+d)$$

The diagram shows that R is the point specified by $(a+c)$ and $(b+d)$ since **PR = OQ** and R the fourth point of the parallelogram OPRQ represents the sum of $(a+ib)$ and $(c+id)$. Clearly addition of complex numbers follows the parallelogram law for vectors. Similarly

$$(a+ib)-(c+id) = (a-c)+i(b-d)$$

which is represented by the point S.

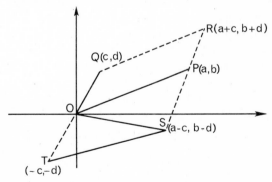

Fig. 131

The multiplication of the complex number $(a+ib)$ by the scalar number s is given by

$$s(a+ib) = sa+isb$$

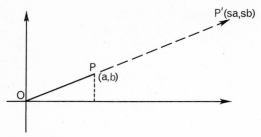

Fig. 132

P represents the complex number $(a+ib)$ and the point P′ with coordinates sa and sb represents $s(a+ib)$. In this way we can give a geometrical interpretation of operations with complex numbers. The whole field of electrical technology and electrical engineering was transformed by the use of the geometrical illustrations of complex number operation which was given by Charles P. Steinwetz in 1893 in a famous paper on 'Complex quantities and their use in Electrical Engineering', read to the International Electrical Congress in Chicago. We can regard the geometrical representation of complex numbers as a *mathematical model*. Other mathematical models are possible such as their representation by (2×2) matrices of the general form $\begin{bmatrix} a & -b \\ b & a \end{bmatrix}$ associated with $a+ib$.

2. The representation of the complex number $a+ib$ by the ordered pair of real numbers $[a, b]$ gives another mathematical model and this system of ordered pairs has the structure of a vector space. The treatment of complex numbers as an ordered pair of real numbers was originated by Sir W. R. Hamilton during his studies which led to the discovery of Quaternions in 1843.

Vectors and matrices

The elements of a matrix can be organised in rows or columns.

Thus the matrix $\begin{pmatrix} 2 & 3 & 1 \\ 4 & 6 & 6 \\ 2 & 1 & 3 \end{pmatrix}$ can be regarded

as three row vectors

$\mathbf{R}_1 = (2, 3, 1)$

$\mathbf{R}_2 = (4, 6, 6)$

$\mathbf{R}_3 = (2, 1, 3)$

or as three volumn vectors

$$\mathbf{C}_1 = \begin{pmatrix} 2 \\ 4 \\ 2 \end{pmatrix}$$

$$\mathbf{C}_2 = \begin{pmatrix} 3 \\ 6 \\ 1 \end{pmatrix}$$

$$\mathbf{C}_3 = \begin{pmatrix} 1 \\ 6 \\ 3 \end{pmatrix}$$

The process of addition of two matrices can then be regarded as the process of adding three pairs of column or row vectors. Thus:

$$\begin{pmatrix} 2 & 3 & 1 \\ 4 & 6 & 6 \\ 2 & 1 & 3 \end{pmatrix} + \begin{pmatrix} 6 & 2 & 1 \\ 3 & 2 & 3 \\ 1 & 6 & 8 \end{pmatrix} = \left(\begin{pmatrix} 2 \\ 4 \\ 2 \end{pmatrix} + \begin{pmatrix} 6 \\ 3 \\ 1 \end{pmatrix} \begin{pmatrix} 3 \\ 6 \\ 1 \end{pmatrix} + \begin{pmatrix} 2 \\ 2 \\ 6 \end{pmatrix} \begin{pmatrix} 1 \\ 6 \\ 3 \end{pmatrix} + \begin{pmatrix} 1 \\ 3 \\ 8 \end{pmatrix} \right)$$

$$= \begin{pmatrix} 8 & 5 & 2 \\ 7 & 8 & 9 \\ 3 & 7 & 11 \end{pmatrix}$$

or using row vectors

$$\begin{pmatrix} 2 & 3 & 1 \\ 4 & 6 & 6 \\ 2 & 1 & 3 \end{pmatrix} + \begin{pmatrix} 6 & 2 & 1 \\ 3 & 2 & 3 \\ 1 & 6 & 8 \end{pmatrix} = \left(\begin{matrix} (2, & 3, & 1) + (6, & 2, & 1) \\ (4, & 6, & 6) + (3, & 2, & 3) \\ (2, & 1, & 3) + (1, & 6, & 8) \end{matrix} \right)$$

$$= \begin{pmatrix} 8 & 5 & 2 \\ 7 & 8 & 9 \\ 3 & 7 & 11 \end{pmatrix}$$

It is often convenient when handling matrices with large numbers of elements, to use a computer and to store the matrices as a number of row or column vectors. The process of multiplication can then be dealt with by using the Scalar Product of the row vectors of the first matrix with the column vectors of the second matrix

$$
\begin{array}{c}
\mathbf{C_1}\ \mathbf{C_2}\ \mathbf{C_3} \\
\begin{matrix}\mathbf{R_1}\\ \mathbf{R_2}\\ \mathbf{R_3}\end{matrix}
\begin{pmatrix} 2 & 3 & 1 \\ 4 & 6 & 6 \\ 2 & 1 & 3 \end{pmatrix}
\times
\begin{pmatrix} 6 & 2 & 1 \\ 3 & 2 & 3 \\ 1 & 6 & 8 \end{pmatrix}
=
\begin{pmatrix} \mathbf{R_1 . C_1} & \mathbf{R_1 . C_2} & \mathbf{R_1 . C_3} \\ \mathbf{R_2 . C_1} & \mathbf{R_2 . C_2} & \mathbf{R_2 . C_3} \\ \mathbf{R_3 . C_1} & \mathbf{R_3 . C_2} & \mathbf{R_3 . C_3} \end{pmatrix}
\end{array}
$$

$$
=
\begin{pmatrix} 22 & 16 & 19 \\ 48 & 56 & 70 \\ 18 & 24 & 29 \end{pmatrix}
$$

With a large matrix, say

$$
\begin{pmatrix}
3 & 2 & 1 & 6 & 4 \\
4 & 3 & 1 & 0 & 2 \\
1 & 0 & 1 & 2 & 1 \\
3 & 6 & 2 & 1 & 3 \\
1 & 1 & 2 & 1 & 1
\end{pmatrix}
$$

each row vector or column vector will have five numbers defining it in their own special order. We regard five ordered numbers as a vector and no longer can we think of magnitude and direction in the sense that was used when the subject of vectors was first introduced. We can illustrate a three-element by a geometrical model in three-dimensional space, we can only think of a five-element vector in an abstract five-dimensional space.

Scalar multiplication of n-vectors
The process of scalar multiplication between two vectors can be extended to n-vectors, so that if $\mathbf{a} = (a_1, a_2, a_3, a_4, a_5 \ldots, a_n)$ and $\mathbf{b} = (b_1, b_2, b_3, b_4, \ldots, b_n)$, then $\mathbf{a} . \mathbf{b}$ is defined as $(a_1 . b_1 + a_2 . b_2 + a_3 . b_3 + \cdots + a_n b_n)$. It is usual to call this process finding the Inner Product when $n > 3$. It is important to emphasise that if $n > 3$ then the vector product is *not* defined.

Group structure
A system of n-vectors has group structure under the operation of addition since:

(i) $\mathbf{A} + \mathbf{B} = \mathbf{C}$ (closure)
(ii) $\mathbf{A} + (\mathbf{B} + \mathbf{C}) = (\mathbf{A} + \mathbf{B}) + \mathbf{c}$ (associativity)
(iii) $\mathbf{A} + \mathbf{0} = \mathbf{A}$ (neutral element for addition)
(iv) $\mathbf{A} + (-\mathbf{A}) = \mathbf{0}$ (inverse for addition)

These axioms establish the group structure; in addition, since $\mathbf{A} + \mathbf{B} = \mathbf{B} + \mathbf{A}$ the group has commutative properties.

The existence of group structure guarantees that there must exist a solution of an equation of the type

$$\mathbf{A} + \mathbf{X} = \mathbf{B}$$

since $[(-\mathbf{A}) + \mathbf{A}] + \mathbf{X} = (-\mathbf{A}) + \mathbf{B}$

$$\therefore \qquad \mathbf{0} + \mathbf{X} = \mathbf{B} + (-\mathbf{A})$$

$$\mathbf{X} = \mathbf{B} - \mathbf{A}$$

It must be emphasised that if $\mathbf{A}$ and $\mathbf{B}$ are 3-vectors then the solution $\mathbf{X}$ must also be a 3-vector. Similarly, if $\mathbf{A}$ and $\mathbf{B}$ are n-vectors then $\mathbf{X}$ must also be an n-vector.

It is beyond the scope of this book to do more than this but the algebra of vector spaces is more usually known as LINEAR ALGEBRA. Linear algebra has wide applications in economics, programming, computing, econometrics, etc. A number of modern books are listed in the bibliography.

Exercise 17.
1. If $\mathbf{s} = (2, 3, 5, 1)$, $\mathbf{t} = (1, 2, 1, 4)$ and $\mathbf{v} = (5, 6, 17, -8)$ express the vector $\mathbf{v}$ as a linear combination of vectors $\mathbf{s}$ and $\mathbf{t}$.
2. Show that the vectors $\mathbf{a} = (3, 2, 1)$, $\mathbf{b} = (4, 2, 2)$ and $\mathbf{c} = (1, 3, 1)$ are linearly independent.
3. What is the solution space of the equation $3x + 4y + 2z = 0$?
4. Find the solution space of the pair of equations

$$3x + 4y + 2z = 20$$
$$2x + 3y + z = 10$$

5. Use the results of Qu. 3 to show that the linear equations

$$3x_1 + 4x_2 + 2x_3 = 20$$
$$2x_1 + 3x_2 + x_3 = 10$$
$$5x_1 + 4x_2 - x_3 = 4$$

have a unique solution. Give the solution vector.

7

6. For what value of m have the following equations a solution?

$3x + 4y + z\ = 11$

$4x + 2y - 2z = 8$

$3x - 2y - 5z = m$

7. Show that numbers of the form $a + b\sqrt{2}$, where a and b are rational, follow the axioms for a vector space.

8. Show that polynomials of the form $ax^3 + bx^2 + cx + d$ also follow the axioms for a vector space.

9. Using the definition of addition and scalar multiplication given on page 142 show that vectors of the form $\mathbf{V} = (a_1, a_2, \ldots a_n)$ follow the axioms for a vector space of n dimensions.

10. Show that position vectors $\mathbf{OP} = x_1\mathbf{i} + y_1\mathbf{j} + z_1\mathbf{k}$ constitute a vector space of three dimensions.

11. In a system of n-vectors under the operation of multiplication by a scalar can we always find a unique solution to equations of the form

$s\mathbf{A} + \mathbf{X} = t\mathbf{B}$

Give reasons for your answer.

12. If matrix $M = \begin{pmatrix} 1 & 2 & 1 & 3 \\ 1 & 0 & 2 & 1 \\ 3 & 3 & 4 & 1 \\ 1 & 0 & 1 & 2 \end{pmatrix}$ and $N = \begin{pmatrix} 1 & 1 & 2 & 1 \\ 1 & 0 & 1 & 2 \\ 3 & 0 & 1 & 4 \\ 1 & 1 & 2 & 3 \end{pmatrix}$

partition matrix M into 4 row vectors and matrix N into 4 column vectors. Using scalar multiplication find $M \cdot N$.

Chapter 12
Vector operators, gradient and field

We can represent vectors in two or three dimensions by directed line segments and the study and analysis of the operations on these directed line segments is usually known as *vector analysis*. This part of the subject was developed to enable engineers and physicists to study problems in mechanics, aerodynamics, electricity, elasticity, relativity, heat and so on, generally in three dimensions. Many of these problems are beyond the scope of this book but some of the fundamental ideas used in tackling them can be introduced and the way prepared for later work. This will also be an appropriate opportunity to link up some past work such as differentiation, integration and partial differentiation by indicating their applications.

It is convenient to use cartesian coordinates but many of the results are independent of the system of reference, i.e. the operations remain the same in spite of a change of axes or origin. Properties or operations which do not change with choice of different axes or points of reference are said to be *invariant* and a consideration of *invariance* is a subject for other books which carry the subject further.

The location of points, lines and surfaces in 3-space requires three variables x, y, z, in cartesian coordinates. The study of curves and surfaces in space is called differential geometry and makes use of the calculus of vectors. Since the subject requires the use of three variables it is inevitable that the treatment involves the use of partial differentiation. Again the applications of vector analysis to scalar fields and vector fields is of considerable importance to the physicist and engineer.

Scalar fields

It is well known that if a hot body is embedded in a block of asbestos, as in night storage heaters, the temperature in the region surrounding the hot body varies with the position of each point under consideration. A similar situation is found in the earth itself. It is established that at a depth of 15 m the temperature of the earth becomes steady at 15·5°C and then increases by 1°C for every 27·4 m nearer the centre of the earth. Temperature is a scalar quantity and these are therefore examples of a scalar field. If in a region of space we can associate a scalar quantity with every point in that space, such a region is called a SCALAR FIELD. If we use cartesian coordinates, and to each point (x, y, z) of the region R in space there can be associated a number or scalar $\phi(x, y, z)$ then ϕ is called a *scalar function* of position or a scalar point-function and this defines a SCALAR FIELD ϕ in region R. If this scalar field is independent of time then it is called a stationary or steady-state scalar field.

Instead of using cartesian coordinates each point P in the region could be determined by its position vector and the scalar function for each point P denoted by $\phi(P)$. As we move from point P to another point P′ in space near to P the scalar function has a new value $\phi(P')$ and the new point P′ lies in the space surrounding P. If the position vector of P is $\mathbf{r}$ and of P′ is $\mathbf{r} + \delta\mathbf{r}$ then the displacement **PP′** is $\delta\mathbf{r}$.

Over the displacement **PP′** the scalar function has changed from $\phi(P)$ to $\phi(P')$.

Then the *average space-rate* of change is

$$\frac{\phi(P') - \phi(P)}{|\delta\mathbf{r}|}$$

and this change takes place in the direction of **PP′** (or $\delta\mathbf{r}$) i.e. it has magnitude and direction. Now as $|\delta\mathbf{r}| \to 0$ the point P′ approaches nearer and nearer to P and *if the limit exists*, then

$$\lim_{|\delta\mathbf{r}| \to 0} \frac{\phi(P') - \phi(P)}{|\delta\mathbf{r}|}$$

represents the *space rate of change* at the point P of the Scalar point function $\phi(P)$ in the direction of $\delta\mathbf{r}$. This *space rate of change* has a magnitude which depends on the direction chosen.

If we denote the space rate of change of ϕ with distance as $(d\phi/ds)$ then if $\hat{\mathbf{u}}$ is the unit vector giving the direction chosen, we show that $(d\phi/ds)$ is in this direction by the subscript $\hat{\mathbf{u}}$ thus $(d\phi/ds)_{\hat{\mathbf{u}}}$ which means the space rate of change in the direction of $\hat{\mathbf{u}}$.

Definition

If $\hat{\mathbf{n}}$ is the unit vector in the direction for which $(d\phi/ds)_{\hat{\mathbf{u}} = \hat{\mathbf{n}}}$ is a maximum then the vector $(d\phi/ds)_{\hat{\mathbf{n}}}\hat{\mathbf{n}}$ is called the *gradient* of the function $\phi(P)$; it is denoted by Grad ϕ or $\nabla\phi$ where the symbol ∇ is usually called DEL (sometimes NABLA). The treatment of Grad ϕ so far has been independent of a system of coordinates, but for

174

convenience it is usually desirable to introduce cartesian coordinates. At the point P the scalar function ϕ is given as $\phi(x, y, z)$. Under these conditions it can be shown that

$$\text{Grad } \phi = \nabla\phi = \frac{\partial\phi}{\partial x}\mathbf{i} + \frac{\partial\phi}{\partial y}\mathbf{j} + \frac{\partial\phi}{\partial z}\mathbf{k}$$

where $\partial\phi/\partial x$, $\partial\phi/\partial y$ and $\partial\phi/\partial z$ are the magnitudes of the components of the vector $\nabla\phi$ along the x, y, z axes. Since ϕ is a function of three independent variables (x, y, z) it is clear that the partial derivatives must be used in the three directions of x, y, z. In the expression for $\nabla\phi$ above, the expression

$$\frac{\partial\phi}{\partial x}\mathbf{i} + \frac{\partial\phi}{\partial y}\mathbf{j} + \frac{\partial\phi}{\partial z}\mathbf{k}$$

can be interpreted as

$$\left(\frac{\partial}{\partial x}\mathbf{i} + \frac{\partial}{\partial y}\mathbf{j} + \frac{\partial}{\partial z}\mathbf{k}\right)\phi$$

treating the expression in the bracket as a sort of VECTOR OPERATOR and regarding this operation as equivalent to ∇ i.e.

$$\nabla \equiv \mathbf{i}\frac{\partial}{\partial x} + \mathbf{j}\frac{\partial}{\partial y} + \mathbf{k}\frac{\partial}{\partial z}$$

Thus at each point P in a scalar field where the scalar function $\phi(x, y, z)$ is differentiable one can associate a vector field given by $\nabla\phi$ such that the *maximum* space rate of change is given in magnitude and direction by this vector field.

For instance if $\phi(x, y, z)$ is a scalar function giving the temperature at each point $P(x, y, z)$ in a field then $\nabla\phi$ gives the direction of the heat flow vector at P.

Example 1

1. If $\phi(x, y, z)$ is a scalar field defined by

$$\phi(x, y, z) = 2x^2y - 3xy^2 + 4xz$$

find the magnitude of the scalar function ϕ at the points (a) $(2, -1, 3)$ and (b) $(-1, 0, 4)$

$$\phi_a = 2(2^2)(-1) - 3(2)(-1)^2 + 4(+2)(3)$$

$$= -8 - 6 + 24$$

$$= +10$$

$$\phi_b = 2(-1)^2 . 0 - 3(-1)0 + 4(-1) . 4$$

$$= 0 + 0 - 16$$

$$= -16$$

Example 2

2. If $\phi(x, y, z) = 2x^2y - 3xy^2 + 4xz$ defines a scalar field find Grad ϕ at the point $(2, -1, 3)$

Grad $\phi = \nabla\phi$	$\phi = 2x^2y - 3xy^2 + 4xz$
$\quad = \mathbf{i}\dfrac{\partial\phi}{\partial x} + \mathbf{j}\dfrac{\partial\phi}{\partial y} + \mathbf{k}\dfrac{\partial\phi}{\partial z}$	$\dfrac{\partial\phi}{\partial x} = 4xy - 3y^2 + 4z = 4 . 2 . (-1)$
$\quad = 1\mathbf{i} + 20\mathbf{j} + 8\mathbf{k}$	$\qquad\qquad -3(-1)^2 + 4 . 3$
	$\qquad\qquad = -8 - 3 + 12 = +1.$
	$\dfrac{\partial\phi}{\partial y} = 2x^2 - 6xy = 2 . 2^2 - 6(2)(-1)$
	$\qquad\qquad = 8 + 12 = 20$
	$\dfrac{\partial\phi}{\partial z} = 4x = 8.$

Hence the maximum space rate of change of function ϕ at the point $(2, -1, 3)$ is $\sqrt{1^2 + 20^2 + 8^2} = \sqrt{465}$ in the direction of the unit vector

$$\hat{\mathbf{r}} = \frac{1}{\sqrt{465}}\mathbf{i} + \frac{20}{\sqrt{465}}\mathbf{j} + \frac{8}{\sqrt{465}}\mathbf{k}$$

Exercise 18.

1. The scalar function $\phi(x, y, z) = x^2 + y^2 + z^2$, find ϕ and $\nabla\phi$ at the point $(1, 2, 3)$.

2. Given a scalar field defined by $\phi(x, y, z) = 2x^2y - zy + 5$ find ϕ at the points (a) $(0, 0, 0)$ (b) $(1, -1, 2)$ (c) $(-1, -1, -2)$.

3. If $\phi(x, y, z) = 2x^2y - zy^2 + 5$ find $\nabla\phi$ (or Grad ϕ) at the point $(1, 1, -2)$.

4. If ϕ_1 and ϕ_2 are differentiable scalar functions of x, y, and z prove that

(i) $\nabla(\phi_1 + \phi_2) = \nabla\phi_1 + \nabla\phi_2$

(ii) $\nabla(\phi_1 . \phi_2) = \phi_2\nabla\phi_1 + \phi_1\nabla\phi_2$

176

The vector operator DEL $\mathbf{V}$ possesses properties which make it useful over a very wide field. It is of course, a vector differential operator since it involves the operations of differentiation and has vector properties also.

We have shown that it can operate directly on a scalar function but if $\mathbf{a}$ is a vector function then $\mathbf{V}a$ is undefined. However, since $\mathbf{V}$ is similar to a vector function scalar products (dot products) and vector products (cross products) are defined and have many applications. The simplest of the scalar products is called the DIRECTIONAL DERIVATIVE obtained by the scalar product of the vector $\mathbf{V}\phi$ and a unit vector $\hat{\mathbf{a}}$ in any required direction.

Directional derivative $= \mathbf{V}\phi \cdot \hat{\mathbf{a}}$

Since Grad ϕ is a vector whose magnitude is the maximum space rate of change at a given point P in space then the scalar product with a unit vector $\hat{\mathbf{a}}$ in any other direction at the same point P gives the magnitude of the space rate of change in the direction of $\hat{\mathbf{a}}$.

Example

Find the space rate of change of the scalar function

$$\phi(x, y, z) = 2x^2 y - 3xy^2 + 4xz$$

at the point $(2, -1, 3)$ in the direction of the vector $2\mathbf{i} + 3\mathbf{j} + 2\mathbf{k}$

$$\text{Grad } \phi \quad \text{or} \quad \mathbf{V}\phi = \mathbf{i}\frac{\partial \phi}{\partial x} + \mathbf{j}\frac{\partial \phi}{\partial y} + \mathbf{k}\frac{\partial \phi}{\partial z}$$

$$= 1\mathbf{i} + 20\mathbf{j} + 8\mathbf{k} \quad \text{(see Example 2 page 154)}$$

If $\mathbf{a} = 2\mathbf{i} + 3\mathbf{j} + 2\mathbf{k}$ then $|\mathbf{a}| = \sqrt{2^2 + 3^2 + 2^2} = \sqrt{17}$

$$\hat{\mathbf{a}} = \frac{\mathbf{a}}{|\mathbf{a}|} = \frac{2}{\sqrt{17}}\mathbf{i} + \frac{3}{\sqrt{17}}\mathbf{j} + \frac{2}{\sqrt{17}}\mathbf{k}$$

The directional derivative of $\mathbf{V}\phi$ with the unit vector $\hat{\mathbf{a}}$ gives the required space rate of change in the direction of $\hat{\mathbf{a}}$

$$\mathbf{V}\phi \cdot \hat{\mathbf{a}} = (1\mathbf{i} + 20\mathbf{j} + 8\mathbf{k}) \cdot \left(\frac{2}{\sqrt{17}}\mathbf{i} + \frac{3}{\sqrt{17}}\mathbf{j} + \frac{2}{\sqrt{17}}\mathbf{k} \right)$$

$$= \frac{2}{\sqrt{17}} + \frac{60}{\sqrt{17}} + \frac{16}{\sqrt{17}}$$

$$= \frac{78}{\sqrt{17}} \simeq 19.$$

Hence required rate $\simeq 19$.

Next consider a scalar function $\phi(x, y, z)$ which is constant (say C)

i.e. $\phi(x, y, z) = C$

The simplest example is the function

$$x^2 + y^2 + z^2 = a^2$$

which gives a *sphere*.
Another example is

$$\frac{x^2}{a^2} + \frac{y^2}{b^2} + \frac{z^2}{c^2} = 1$$

which gives an ellipsoid.

In both examples the scalar function represents a surface in three dimensional space; in general if $\phi(x, y, z) = C$ the function represents a surface.

Let the position vector of a point P on such a surface be **r** then

$$\mathbf{r} = x\mathbf{i} + y\mathbf{j} + z\mathbf{k}$$

and

$$\frac{d\mathbf{r}}{ds} = \mathbf{i}\frac{dx}{ds} + \mathbf{j}\frac{dy}{ds} + \mathbf{k}\frac{dz}{ds}$$

where s is the distance of P from some fixed point. Reference to Fig. 124 shows that $d\mathbf{r}/ds = \hat{\mathbf{T}}$ (a unit vector along the tangent).

and hence $d\mathbf{r}/ds$ lies in the tangent plane to the surface at (x, y, z).

If we take the scalar product of Grad ϕ with $d\mathbf{r}/ds$ we have

$$(\nabla\phi) \cdot \frac{d\mathbf{r}}{ds} = \left(\mathbf{i}\frac{\partial\phi}{\partial x} + \mathbf{j}\frac{\partial\phi}{\partial y} + \mathbf{k}\frac{\partial\phi}{\partial z}\right) \cdot \left(\mathbf{i}\frac{dx}{ds} + \mathbf{j}\frac{dy}{ds} + \mathbf{k}\frac{dz}{ds}\right)$$

$$= \frac{\partial\phi}{\partial x}\frac{dx}{ds} + \frac{\partial\phi}{\partial y}\frac{dy}{ds} + \frac{\partial\phi}{\partial z}\frac{dz}{ds}$$

$$= \frac{d\phi}{ds}$$

where $d\phi/ds$ is the total differential coefficient of $\phi(x, y, z)$ with respect to s.

But $\phi(x, y, z) = \text{constant}$

hence $\dfrac{d\phi}{ds} = 0$

178

This means that

$$(\nabla\phi) \cdot \frac{d\mathbf{r}}{ds} = 0$$

and it follows that the vector $\nabla\phi$ is at right angles to $d\mathbf{r}/ds$ which lies in the tangent plane.

Thus $\nabla\phi$ must be normal to the tangent plane, at the point $P(x, y, z)$.

Example 1
To find the normal to the surface of the ellipsoid

$$\frac{x^2}{2} + \frac{y^2}{4} + \frac{z^2}{9} = 1$$

at the point $(1, 1, \frac{3}{2})$.

The normal to the surface is $\nabla\phi$ at $(1, 1, \frac{3}{2})$

$$= \left(x\mathbf{i} + \frac{y}{2}\mathbf{j} + \frac{2z}{9}\mathbf{k} \right)_{(1,1,\frac{3}{2})}$$

$$= 1\mathbf{i} + \tfrac{1}{2}\mathbf{j} + \tfrac{1}{3}\mathbf{k}.$$

Example 2
To find the equation of the tangent plane to the surface of the ellipsoid $x^2/2 + y^2/4 + z^2/9 = 1$ at the point $(1, 1, \frac{3}{2})$.

The normal to the surface i.e. to the tangent plane at the point $(1, 1, \frac{3}{2})$ has been calculated and is

$$\mathbf{n} = 1\mathbf{i} + \tfrac{1}{2}\mathbf{j} + \tfrac{1}{3}\mathbf{k}$$

If the point (x, y, z) is any point on this plane then its position vector is

$$\mathbf{r} = x\mathbf{i} + y\mathbf{j} + z\mathbf{k}$$

and the position vector of the point $(1, 1, \frac{3}{2})$

$$\mathbf{r}_0 = 1\mathbf{i} + 1\mathbf{j} + \tfrac{3}{2}\mathbf{k}$$

hence

$$(\mathbf{r} - \mathbf{r}_0) \cdot \mathbf{n} = 0$$

gives the vector equation of the required plane. Hence

$$[x\mathbf{i} + y\mathbf{j} + z\mathbf{k} - (1\mathbf{i} + 1\mathbf{j} + \tfrac{3}{2}\mathbf{k})] \cdot [1\mathbf{i} + \tfrac{1}{2}\mathbf{j} + \tfrac{1}{3}\mathbf{k}] = 0$$

$$[(x-1)\mathbf{i} + (y-1)\mathbf{j} + (z-\tfrac{3}{2})\mathbf{k}] \cdot [1\mathbf{i} + \tfrac{1}{2}\mathbf{j} + \tfrac{1}{3}\mathbf{k}] = 0$$

7*

or

$$x - 1 + \tfrac{1}{2}(y - 1) + \tfrac{1}{3}(z - \tfrac{3}{2}) = 0$$

which reduces to

$$6x - 6 + 3y - 3 + 2z - 3 = 0$$

and

$$6x + 3y + 2z - 12 = 0$$

is the required equation of the tangent plane at the point $(1, 1, \tfrac{3}{2})$.

Vector fields

If at any point (x, y, z) in a region we can associate a vector function $\mathbf{a}(x, y, z)$, such a region is called a vector field.

$\mathbf{a}$ is a vector function of position or a vector point function.

A vector field which is independent of time is called a *stationary* or *steady-state* vector field.

Common examples of vector fields are:

1. The magnetic field in the space surrounding a bar magnet; at each point in the space round the magnet the magnetic field has a definite magnitude and direction which vary with the position of the point.
2. The gravitational force field round the earth.
3. The electrostatic force field round a charged body.
4. The velocity at different points in a moving fluid.

If $\mathbf{a}$ is vector function of position we have already noted that $\nabla \mathbf{a}$ is undefined but the vector differential operator ∇ can operate by a scalar process or vector process denoted $\nabla .$ or $\nabla \times$ on another vector.

$\nabla . \mathbf{a}$ is defined as the DIVERGENCE of vector $\mathbf{a}$ or Div. $\mathbf{a}$ (a scalar quantity)

and

$\nabla \times \mathbf{a}$ is defined as the CURL of vector $\mathbf{a}$ (a vector quantity).

It is beyond the scope of this book to pursue this part of the subject further but the scalar field given by div. $\mathbf{a}$ has applications in physics and mechanics.

Exercise 19.

1. The equation $x^2 + y^2 + z^2 = 29$ represents the surface of a sphere, centre at the origin and radius $\sqrt{29}$ units.

(i) Show that the point P(2, 3, 4) lies on the sphere.

(ii) Find the position vector **r** of P.

(iii) Find $\nabla\phi$ at the point P(2, 3, 4) hence verify that $\nabla\phi$ is normal to the surface at P.

(iv) Find the equation of the plane tangential to the sphere at P(2, 3, 4).

2. Find the unit normal vector to the surface $x^2y + 2yz^2 = 6$ at the point (1, 2, 1).

3. Find the equation of the tangent plane to the surface $xy + yz^2 + xz = 11$ at the point (2, 3, 1).

4. Find the angle between the surfaces $x^2 + y^2 + z^2 = 6$ and $x^2 + y^2 - z = 0$ at the point (1, 1, 2) where they cut. (Hint: the angle between two surfaces is the angle between the normals to the surfaces).

5. Defining $\nabla^2\phi$ as $\nabla . \nabla\phi$, deduce that

$$\nabla^2\phi = \frac{\partial^2\phi}{\partial x^2} + \frac{\partial^2\phi}{\partial y^2} + \frac{\partial^2\phi}{\partial z^2}$$

6. If the vector function $\mathbf{a} = a_x\mathbf{i} + a_y\mathbf{j} + a_z\mathbf{k}$, show that

$$\nabla . \mathbf{a} = \frac{\partial(a_x)}{\partial x} + \frac{\partial(a_y)}{\partial y} + \frac{\partial(a_z)}{\partial z}$$

Exercise 20 (Miscellaneous)

1. Show that, when multiplied by a scalar number, vectors in three-dimensional space have a group structure. Show also that the relation 'is parallel to' is an equivalence relation over the same set of vectors.

2. Show that the set of abstract vectors of the form $\mathbf{r} = (x, y, z, t)$ can be transformed to a new basis (x', y, z, t) where $x' = x + vt$, and give the matrix required to perform the transformation.

3. A satellite travelling in space is tracked from a station s on the earth. Radar gives the magnitude of the position vector **SP** as 205000 km and after an interval of time t hours **SP'**, the new position vector, is given as magnitude 205500 km and angle P'SP = 0·0175 rad. If t is 3 hours and the movement of the earth is neglected, what do you estimate the apparent speed of the satellite to be relative to the earth?

4. Plane P_1 contains vectors $\mathbf{a} = a_1\mathbf{i} + b_1\mathbf{j} + c_1\mathbf{k}$ and

$$\mathbf{b} = a_2\mathbf{i} + b_2\mathbf{j} + c_2\mathbf{k}$$

Plane P_2 contains vectors $\mathbf{c} = l_1\mathbf{i} + m_1\mathbf{j} + n\mathbf{k}$
$$\mathbf{d} = l_2\mathbf{i} + m_2\mathbf{j} + n_2\mathbf{k}$$

Under what conditions are the planes P_1 and P_2:
(i) Parallel?
(ii) Perpendicular?
(iii) Neither parallel nor perpendicular?

5. A satellite weighing M kg has a velocity vector $\mathbf{V}_1$, which is at right-angles to its position vector $\mathbf{r}$, measured from the centre of the sun as origin; distances being measured in kilometres and time in seconds. A rocket motor exerts a thrust of F newtons in the positive direction of $\mathbf{r}$. What is the force vector and the velocity vector $\mathbf{V}_F$ after 10 minutes assuming that $\mathbf{r}$ does not alter very appreciably?

6. The direction cosines of the line L,

$$\frac{x-2}{2} = \frac{y-3}{3} = \frac{z-5}{4}$$

are l, m, and n. Give the direction vector of L, find l, m, and n and find $l^2 + m^2 + n^2$.

7. The equation of the surface of a sphere is $x^2 + y^2 + z^2 = 50$. Show that the point $P(3, 4, 5)$ lies on the sphere. Find the equation of the tangent plane through P.

8. If a plane undergoes a translation given by the shift vector $\begin{pmatrix} a \\ b \end{pmatrix}$ and the position vector of a point P is $\mathbf{p} = \begin{pmatrix} x \\ y \end{pmatrix}$ before translation, and $P' = \begin{pmatrix} x' \\ y' \end{pmatrix}$ is the position vector afterwards, give the matrix M of the translation.

9. The line

$$\frac{x-2}{2} = \frac{y-3}{3} = \frac{z-5}{4}$$

passes through the point $(4, 6, 9)$, which also lies in the plane $2x + 3y - 3z + 1 = 0$. What is the angle between the line and the plane?

10. Given that $\mathbf{a} = 2\mathbf{i} + 3\mathbf{j}$, $\mathbf{b} = 1\mathbf{i} - 2\mathbf{j}$, $\mathbf{c} = 2\mathbf{j} + 3\mathbf{k}$, can you find scalars l, m, and n, such that

$$l\mathbf{a} + m\mathbf{b} + n\mathbf{c} = 0$$

What name is given to such a set of vectors as **a**, **b**, and **c**?

If $\mathbf{V} = 3\mathbf{i} + 7\mathbf{j} + 9\mathbf{k}$

what values of x, y, and z will make $x\mathbf{a} + y\mathbf{b} + z\mathbf{c} = \mathbf{V}$?
What name is given to such a set of vectors as **a**, **b**, **c**, and **v**?

11. Find the coordinates of a point U, such that the position vector of U is of unit length and parallel to the line joining P(3, 3, 4) to Q(5, 4, 6). What are the direction cosines of the line PQ? What is the equation of the plane defined by the three points P, Q, and U?

12. Simplify: $(\mathbf{a} + \mathbf{b}) . (\mathbf{a} - \mathbf{b})$

13. Simplify: $(\mathbf{a} + \mathbf{b}) \times \mathbf{a} - \mathbf{b})$

14. Simplify: (i) $(\mathbf{a} + \mathbf{b}) . (\mathbf{a} + \mathbf{b})$; (ii) $(\mathbf{a} + \mathbf{b}) \times (\mathbf{a} + \mathbf{b})$

15. Simplify: $(\mathbf{a} + \mathbf{b}) \times (\mathbf{a} + \mathbf{c})$

16. If **a** and **b** and **c** are radius vectors of the same circle,

 (i) what can you say about the vectors $(\mathbf{a} + \mathbf{b})$ and $(\mathbf{a} - \mathbf{b})$?

 (ii) if $\mathbf{d} = (\mathbf{a} + \mathbf{b}) \times (\mathbf{a} - \mathbf{b})$, what is the value of **d**?

 (iii) if $\mathbf{e} = (\mathbf{a} + \mathbf{b}) \times (\mathbf{a} + \mathbf{c})$, what is the minimum of **e**?

 (iv) What is the value of $(\mathbf{a} + \mathbf{b}) . (\mathbf{a} - \mathbf{b})$?

17. In a rectangular room, 6 m by 30 m and 3 m high, a point 2 m from the floor and at the centre of one of the long walls is taken as origin. The position vector **r** of a small insect buzzing about is given by

$$\mathbf{r} = (3 + \cos pt)\mathbf{i} + (5 \sin pt)\mathbf{j} + (2 + \sin 7pt)\mathbf{k}$$

If $p = \pi/4$ radians, and **i** is a unit vector towards the mid-point of the opposite wall;

 (i) what is the position vector after 4 secs?

 (ii) what is the velocity vector after 4 secs?

 (iii) at what time will the insect next be back in exactly the same position from which it started (i.e., when $t = 0$)?

 (iv) at what times will the magnitude of the position vector be a minimum?

 (v) at what times will the insect be nearest to the ceiling?

18. An aircraft has a speed of approximately 1000 km/hr. Taking the x-axis in an Easterly direction and the z-axis vertically upwards, give the velocity vector $\mathbf{v}_1$ of the aircraft if it climbs steadily in a North-easterly direction and increasing its height by 4 km/hr.

If a second aircraft leaves from the same point with a velocity vector $\mathbf{v}_2 = 500\mathbf{i} + 8\mathbf{k}$, what is the relative velocity of the second aircraft to the first?

How far apart (to nearest kilometre) are they after 2 hours?

19. Aircraft A has a constant velocity vector v_1 and aircraft B has a constant velocity v_2.
   (i) What is the velocity of B relative to A if they start their flights from the same point O?
   (ii) What will be their position vectors p_1 and p_2 after a time t_1, taking O as origin?
   (iii) What will be their distance apart after time t_1?
   (iv) What must be the new velocity vector v_3 of aircraft A, if they are to meet after a further time t_2?

20. A particle P is at a point B at time $t = 0$, and is moving with a speed s, which is related to the time t by the formula $s = kt + s'$ and is in the direction of a constant vector a. Give the position vector r of P from an origin 0 after t secs. What is the velocity vector at time t and the acceleration vector? (The unit of length is the metre.)

Bibliography

IRVING ADLER. *The New Mathematics*. Dobson, 1959; Cygnet Paper Book.

ASSOCIATION OF TEACHERS OF MATHEMATICS. *Some Lessons in Mathematics*. Cambridge University Press, 1964.

G. BIRCHOFF and S. MACLANE. *A Survey of Modern Algebra*. Macmillan, 1953.

P. M. COHN. *Linear Equations*. Routledge & Kegan Paul, 1960.

C. J. ELIEZER. *Concise Vector Analysis*. Pergamon Press.

V. M. FADEEVA. *Computational Methods of Linear Algebra*. Dover Publications, 1959.

J. W. GIBBS. *The Scientific Papers of J. Willard Gibbs*. 2 vols. Longmans, 1905. Vol. 2.

G. E. HAY. *Vector and Tensor Analysis*. Dover Publications.

A. M. MACBEATH. *Elementary Vector Algebra*. Oxford University Press, 1964.

MATHEMATICAL ASSOCIATION. *A Second Report on the Teaching of Mechanics in Schools*. Bell, 1965.

D. C. MURDOCH. *Linear Algebra for Undergraduates*. Wiley, 1957.

M. M. NICHOLSON. *Mathematics for Science Students*. Longmans, 1961.

O.E.E.C. *New Thinking in School Mathematics*. Paris, 1961.

O.E.E.C. *Synopses for Modern Secondary School Mathematics*. Paris, 1961.

D. E. RUTHERFORD. *Vector Methods*. Oliver & Boyd, 1946.

W. W. SAWYER. *Concrete Approach to Abstract Algebra*. Freeman, 1959.

S. SCHUSTER. *Elementary Vector Geometry*. J. Wiley, 1962.

B. SPAIN. *Vector Analysis*. Van Nostrand, 1965.

M. R. SPIEGEL. *Theory and Problems of Vector Analysis*. Schaum, 1959.

G. STEPHENSON. *Mathematics for Science Students*. Longmans, 1961.

C. E. WEATHERBURN. *Elementary Vector Analysis*. Bell (1965 edition).

E. WOLSTENHOLME. *Elementary Vectors*. Pergamon Press, 1964.

Answers

Exercise 1.

1. Speed 260 km/h direction N 40° 53′ W (or N 0·7135 rad W).

2. 17·7 km/h bearing 121° 17′ (or 2·1168 rad).

3. 10·4 N bisecting angle between the two forces.

4. 9·85 m/s angle 5° 22′ (or 0·0937 rad) with the vertical.

5. 24·49 km/h TOWARDS N 67° 17′ W (or N 1·1743 rad W).

6. At point P 5·928 km N and 2·828 km E, i.e. 6·48 km in a direction N 25° 53′ E (or N 0·4517 rad E) from O. Conclusion addition of vectors is commutative.

Exercise 2.

1. 8·9 km/h   S 82° 23′ W (1·4379 rad).

2. 64 km/h from N 51° 20′ W (0·8959 rad).

3. 19·2 km/h from N 6° 20′ E (0·1105 rad).

4. 43·24 km/h from N 19° 6′ E (0·3333 rad).

5. 32·6 km/h from N 85° 36′ E (1·4940 rad).

Exercise 3.

1. $|\mathbf{c}| = 4\cdot978$ 41° 30′ with OX.

2. $|\mathbf{d}| = 14\cdot934$ 41° 30′ with OX.

3. $\mathbf{a} = 3\hat{\mathbf{a}}$

4. $\mathbf{b} = 2\hat{\mathbf{b}}$

5. $\mathbf{c} = 4\cdot978\hat{\mathbf{c}}$

   $|\mathbf{d}| = 14\cdot934$ $\mathbf{d} = 14\cdot934\hat{\mathbf{c}}$

6. Resultant $= 3\cdot606$ N   S 76° 51′ E (1·3413 rad).

Exercise 4.

1. 6·06 N making an angle of 30° (0·5236 rad) between the forces of 5 N and 4 N.

Equilibriant force 6·06 N acting at P in the same plane, and in a direction bisecting angle between forces 2 N and $3\frac{1}{2}$ N.

2. $x = \frac{3}{2}$ and $y = \frac{3}{2}$.

4. $\mathbf{c} = 2\mathbf{d}$ PR$\parallel$ST, PR $= 2$ST. Mid-point theorem.

5. Two sides and the included angle.

Exercise 4a.

1. $7\,\mathbf{AD}$

3. $R = 6\sqrt{37}$ N acting along OD where D is the point (6, 1).

5. $2\mathbf{AQ} = \mathbf{a} - \mathbf{c}$, $2\mathbf{BR} = \mathbf{b} - \mathbf{a}$, $2\mathbf{CP} = \mathbf{c} - \mathbf{b}$

7. $\text{AE}/\text{EC} = \frac{1}{2}$

10. Force: Magnitude $= 5$AD, acting along AD where D cuts BC in the ratio of $2:3$.

Exercise 5.

1. $\mathbf{p} = 3\hat{\mathbf{x}} + 4\hat{\mathbf{y}}$; $\mathbf{q} = 5\hat{\mathbf{x}} + 3\hat{\mathbf{y}}$; $\mathbf{PQ} = \mathbf{q} - \mathbf{p} = 2\hat{\mathbf{x}} - \hat{\mathbf{y}}$

2. $\mathbf{OM} = 4\hat{\mathbf{x}} + \frac{7}{2}\hat{\mathbf{y}}$

3. $\mathbf{OG} = 5\hat{\mathbf{x}} + 5\hat{\mathbf{y}}$

4. $\mathbf{AB} = 2\hat{\mathbf{x}} + \hat{\mathbf{y}}$ $\mathbf{BC} = 2\hat{\mathbf{x}} + \hat{\mathbf{y}}$. These are parallel vectors with a common point B, hence collinear.

8. $\mathbf{AC} = \mathbf{a} + \mathbf{b}$; $\mathbf{BD} = \mathbf{b} + \mathbf{c}$; $\mathbf{AD} = (\mathbf{a} + \mathbf{b}) + \mathbf{c}$ and $\mathbf{AD} = \mathbf{a} + (\mathbf{b} + \mathbf{c})$. Addition is associative.

10. $\mathbf{c} = \mathbf{b} - \mathbf{a}$; $\mathbf{EF} = m\mathbf{b} - m\mathbf{a} = m(\mathbf{b} - \mathbf{a}) = m\mathbf{c}$. $|\mathbf{EF}| = m|\mathbf{c}|$. Corresponding sides are in the same ratio and are parallel. Triangles are similar.

11. $\mathbf{OG} = 5\hat{\mathbf{x}} + 5\hat{\mathbf{y}}$. G is the point (5, 5).

12. $\mathbf{OD} = \dfrac{11\hat{\mathbf{x}}}{3} + 4\hat{\mathbf{y}}$

14. $\mathbf{OG} = -\dfrac{(\mathbf{a} + \mathbf{b})}{10}$

G lies on OM where M is the mid-point of DC and $\mathbf{OG} = \frac{1}{5}\mathbf{OM}$.

An Introduction to Vectors

Exercise 6.

1. $|\mathbf{a}| = 10$

   $\hat{\mathbf{a}} = 0{\cdot}6\hat{\mathbf{x}} + 0{\cdot}8\hat{\mathbf{y}}$

2. $\mathbf{a} = 3\hat{\mathbf{x}} + 4\hat{\mathbf{y}}$

   $\mathbf{b} = 6\hat{\mathbf{x}} + 7\hat{\mathbf{y}}$

   $\mathbf{AB} = 3\hat{\mathbf{x}} + 3\hat{\mathbf{y}} = \begin{bmatrix} 3 \\ 3 \end{bmatrix}$

3. $\mathbf{AB} = 3\hat{\mathbf{x}} + 3\hat{\mathbf{y}}; \ \mathbf{AP} = 5\hat{\mathbf{x}} + 5\hat{\mathbf{y}}$

   Since $\mathbf{AP} = \tfrac{5}{3}\mathbf{AB}$ they are parallel vectors having point A in common, hence the points A, B and P are collinear.

4. $\mathbf{AB} = 2\hat{\mathbf{x}} + 2\hat{\mathbf{y}}$

   $\mathbf{BC} = -1\hat{\mathbf{x}} + 2\hat{\mathbf{y}}$

   $\mathbf{AC} = 1\hat{\mathbf{x}} + 4\hat{\mathbf{y}}$

   $\mathbf{AB} + \mathbf{BC} = \hat{\mathbf{x}} + 4\hat{\mathbf{y}}$

   $\mathbf{OG} = 3\hat{\mathbf{x}} + 5\hat{\mathbf{y}}$

5. $\mathbf{AB} = \dfrac{10\sqrt{13}}{13}\hat{\mathbf{x}} + \dfrac{15\sqrt{13}}{13}\hat{\mathbf{y}}$

6. $\mathbf{a} = -3\hat{\mathbf{x}} - 2\hat{\mathbf{y}}$

   $\mathbf{b} = 18\hat{\mathbf{x}} + 12\hat{\mathbf{y}}$

   $\mathbf{c} = -3k\hat{\mathbf{x}} - 2k\hat{\mathbf{y}}$

7. $\mathbf{u}_1 = \dfrac{5\sqrt{26}}{26}\hat{\mathbf{x}} + \dfrac{\sqrt{26}}{26}\hat{\mathbf{y}}$

   $\mathbf{u}_2 = -\dfrac{\sqrt{26}}{26}\hat{\mathbf{x}} + \dfrac{5\sqrt{26}}{26}\hat{\mathbf{y}}$

   $\mathbf{u}_3 = \hat{\mathbf{x}}$

Exercise 7.

1. $\mathbf{a} = 3\mathbf{b} - 2\mathbf{c}$

2. $\mathbf{r} = -\dfrac{13\hat{\mathbf{x}}}{2} - 6\hat{\mathbf{y}}$

3. $m = -12 \qquad n = 12$

4. In questions 1, 2 and 3, linear dependence has been shown algebraically. In this question the linear dependence of **c** on **a** and **b** is shown geometrically.

$$c = \frac{4}{3}a + \frac{5}{4}b$$

5. (i) $\dfrac{b}{|b|}$ (ii) $\dfrac{c}{|c|}$ (iii) bisects $\widehat{BOC}$

6. $AB = c \qquad AC = c - a$

$BC = -a \qquad OB = c + a$ perpendicular

7. $q = \begin{pmatrix} 5 \\ 4 \end{pmatrix}$. Collinear.

Exercise 8.

9. $m = -3$

10. $\theta = \cos^{-1} \frac{16}{65}$

Exercise 9.

1. $4i - j + k$. 8. 1. 9. Conclusion these results verify that $a \cdot (b + c) = a \cdot b + a \cdot c$.

2. 39. Angle is arc $\cos \dfrac{39\sqrt{61}}{305}$

3. $i + 3j - 2k$. 21. 19. These results verify that $r \cdot (p - q) = r \cdot p - r \cdot q$.

5. $r \cdot (p + q) = 40$. $(r + p) \cdot q = 44$.

6. (i) $4i - j + 6k$ (ii) $7i + j + 2k$

   (iii) $5i - k$ (iv) $7i + j + 2k$

   (v) $22i + 7i - k$

7. $\sqrt{14}$; $\sqrt{17}$; $\sqrt{29}$; $\sqrt{53}$.

8. arc $\cos \dfrac{11\sqrt{238}}{238}$ arc $\cos \dfrac{(-10\sqrt{493})}{493}$

9. $p = 161/51, \quad q = -62/51, \quad r = 36/51$

189

10. $\dfrac{\sqrt{53}}{53}(4\mathbf{i}-\mathbf{j}+6\mathbf{k})$

11. $\mathbf{a} = \begin{bmatrix} 2 \\ 1 \\ 3 \end{bmatrix}$ $\mathbf{b} = \begin{bmatrix} 2 \\ -2 \\ 3 \end{bmatrix}$ $\mathbf{c} = \begin{bmatrix} 3 \\ 2 \\ -4 \end{bmatrix}$

(i) $\begin{bmatrix} -4 \\ -1 \\ 6 \end{bmatrix}$ (ii) $\begin{bmatrix} 7 \\ 1 \\ 2 \end{bmatrix}$ (iii) $\begin{bmatrix} 5 \\ 0 \\ -1 \end{bmatrix}$ (iv) $\begin{bmatrix} 7 \\ 1 \\ 2 \end{bmatrix}$ (v) $\begin{bmatrix} 22 \\ 7 \\ -1 \end{bmatrix}$

12. $\dfrac{22\sqrt{13}}{13}$

13. $20/3$

14. $\dfrac{106\sqrt{17}}{17}$

16. $\cos\theta_1 = \dfrac{x_1}{\sqrt{x_1^2 + y_1^2 + z_1^2}}$

$\cos\theta_2 = \dfrac{y_1}{\sqrt{x_1^2 + y_1^2 + z_1^2}}$

$\cos\theta_3 = \dfrac{z_1}{\sqrt{x_1^2 + y_1^2 + z_1^2}}$

18. $\mathbf{n} = -b\mathbf{i} + a\mathbf{j}$

or $\mathbf{n} = b\mathbf{i} - a\mathbf{j}$

19. $\mathbf{m} = -3b\mathbf{i} + 3a\mathbf{j}$

or $\mathbf{m} = 3b\mathbf{i} - 3a\mathbf{j}$

20. $\mathbf{s} = n(2\sqrt{6}\mathbf{i} - 3\mathbf{k})$

Exercise 10.

1. $\mathbf{r} = 2a\cos^2\theta\mathbf{i} + a\sin 2\theta\mathbf{j}$

2. $\mathbf{r} = a\sin 2\theta\mathbf{i} + 2a\sin^2\theta\mathbf{j}$

3. $\mathbf{r} = a\cos\theta\mathbf{i} + b\sin\theta\mathbf{j}$

4. $r = \dfrac{2p \cos \theta}{1 - \cos \theta} \mathbf{i} + \dfrac{2p \sin \theta}{1 - \cos \theta} \mathbf{j}$

5. Rectangular hyperbola. $xy = c^2$

6. Straight line parallel to vector $\mathbf{p}$

7. $\mathbf{r} = a \sin \dfrac{t}{2} \mathbf{i} + a \cos \dfrac{t}{2} \mathbf{j} + \dfrac{a(8\pi - t)}{2} \mathbf{k}$

8. (i) $\mathbf{r} = \sqrt{3a} \sin \theta \mathbf{i} + \sqrt{3a} \cos \theta \mathbf{j} + (1 + \cos \theta) \mathbf{k}$

   (ii) Ellipse

9. Hyperbola

$$\dfrac{x^2}{a^2} - \dfrac{(z-h)^2}{b^2} = 1$$

10. (i) Elliptical S.H.M.   (ii) Circular Motion   (iii) Linear S.H.M.

11. $\mathbf{r} = at^2 \mathbf{i} + 2at \mathbf{j}$

12. Parabolic.

Exercise 11.

1. $|\mathbf{a} \times \mathbf{b}| = |\mathbf{a}| \, |\mathbf{b}| \sin 30° = 6 \text{ units.}$

Vector product is a vector $6\hat{\mathbf{u}}$ at right angles to the plane of $\mathbf{a}$ and $\mathbf{b}$.

2. $(x_1 y_2 - y_1 x_2) \hat{\mathbf{z}}$

3. $\mathbf{a} \times \mathbf{b} = 4\mathbf{k}.$ $\mathbf{a} \times \mathbf{b} = |\mathbf{a}| \, |\mathbf{b}| \sin \theta \hat{\mathbf{u}}.$ $\hat{\mathbf{u}} = \mathbf{k}$

$5\sqrt{20} \sin \theta \, \mathbf{k} = 4\mathbf{k}$

$\theta = 10° \, 18' \left(\text{or, arc } \sin \theta = \dfrac{2\sqrt{5}}{25} \right)$

6. $\mathbf{a}$ and $\mathbf{b}$ are parallel or collinear.

7. $\dfrac{m}{2} = \dfrac{n}{3} = \dfrac{12}{5}$

8. $\mathbf{r} \cdot \mathbf{s} = 25.$ $\mathbf{r} \times \mathbf{s} = -4\mathbf{i} + \mathbf{j} + 2\mathbf{k}$

First product is a scalar and the second is a vector.

9. $|\mathbf{m} \times \mathbf{n}| = 10\sqrt{3}.$ The vector product is $10\sqrt{3}\hat{\mathbf{u}}$ at right-angles to the plane of $\mathbf{m}$ and $\mathbf{n}$.

An Introduction to Vectors

10. $|\mathbf{m}\times\mathbf{n}| = -18\mathbf{i}+18\mathbf{j}+0\mathbf{k}$

   $\hat{\mathbf{u}} = \dfrac{\sqrt{2}}{2}(-1\mathbf{i}+1\mathbf{j}+0\mathbf{k})$

13. $(\mathbf{a}\times\mathbf{b})$ is a vector perpendicular to both $\mathbf{a}$ and $\mathbf{b}$.

Exercise 11a.
1. (i) $(x-p)\mathbf{i}+(y-q)\mathbf{j}+(z-r)\mathbf{k} = t(a\mathbf{i}+b\mathbf{j}+c\mathbf{k})$

   (ii) $\dfrac{x-p}{a} = \dfrac{y-q}{b} = \dfrac{z-r}{c} (= t)$

2. $\frac{1}{7}(2\mathbf{i}+3\mathbf{j}+6\mathbf{k})$

3. $d = 2 \quad a = 2$
   $e = 3 \quad b = 1$
   $f = 6 \quad c = -2$

4. Yes, they are the same straight line.

5. $\dfrac{x}{2} = \dfrac{y-1}{3} = \dfrac{z-3}{6}$

   $3\mathbf{i}+3\mathbf{j}+4\mathbf{k}$

   $\dfrac{1}{\sqrt{34}}(3\mathbf{i}+3\mathbf{j}+4\mathbf{k})$

   $\dfrac{\sqrt{34}}{2}$

Exercise 11b.
1. $\mathbf{d} = 1\mathbf{i}+2\mathbf{j}+3\mathbf{k}$

2. $\hat{\mathbf{d}} = \dfrac{1}{\sqrt{14}}(1\mathbf{i}+2\mathbf{j}+3\mathbf{k})$

3. (i) Yes (ii) Same line

4. $\dfrac{1}{\sqrt{14}}, \dfrac{2}{\sqrt{14}}, \dfrac{3}{\sqrt{14}}$

   or $\dfrac{\sqrt{14}}{14}, \dfrac{\sqrt{14}}{7}, \dfrac{3\sqrt{14}}{14}$

5. $p\mathbf{i} + q\mathbf{j} + r\mathbf{k}$

6. Parallel but not coincident.

7. $\dfrac{13\sqrt{14}}{14}$

8. $2\mathbf{i} + 5\mathbf{j} - 4k$

9. (i) No
    (ii) Skew
    (iii) $\dfrac{2}{\sqrt{45}} = \dfrac{2\sqrt{45}}{45}$

10. The plane $2x + 3y - 6z + 4 = 0$.

Exercise 12.

1. $\frac{1}{2}|(\mathbf{r}_2 - \mathbf{r}_1) \times (\mathbf{r}_3 - \mathbf{r}_2)|$

2. $(\mathbf{r}_2 - \mathbf{r}_1) \times (\mathbf{r}_3 - \mathbf{r}_2) = \mathbf{0}$

3. $(\mathbf{r}_2 - \mathbf{r}_1) . (\mathbf{r}_3 - \mathbf{r}_2) = 0$

4. $7x - 4y - z + 1 = 0$

5. $\dfrac{7\mathbf{i} - 4\mathbf{j} - \mathbf{k}}{\sqrt{66}}$

6. $\dfrac{\sqrt{66}}{2}$

7. $\mathbf{r} = (1 - t)\mathbf{a} + t\mathbf{b}$

8. $\dfrac{x - 2}{3} = \dfrac{y - 3}{4} = \dfrac{z - 3}{5}$

9. $\dfrac{9\sqrt{46}}{46}$

13. $xx' + yy' + zz' - a^2 = 0$

14. $\dfrac{x - 10\frac{1}{2}}{-4\frac{1}{2}} = \dfrac{y}{6} = \dfrac{z}{8}$

15. $3\mathbf{i} + 2\mathbf{j} + 6\mathbf{k}$. $d = 4$ units.

16. They are all parallel to the plane $2x + 3y + 6z = 0$.

17. $\dfrac{x+11\frac{1}{2}}{18} = \dfrac{y-1\frac{3}{4}}{-3} = \dfrac{z-8}{-8}$

18. $(x-h)^2 + (y-k)^2 + (z-l)^2 = r^2$

19. $(a-h)x + (b-k)y + (c-l)z + k = 0$

    $k = [a(a-h) + b(b-k) + c(c-l)]$

20. $\mathbf{v} = \dfrac{|\mathbf{r}|\mathbf{i}}{\sqrt{2}} + \dfrac{|\mathbf{r}|\mathbf{j}}{\sqrt{2}} + 0\mathbf{k}$

21. $\dfrac{5\sqrt{2}y}{2} + \dfrac{5\sqrt{2}z}{2} + k = 0$

    where $k = \dfrac{-5(10 + 11\sqrt{2})}{2}$

Exercise 13.

1. $\mathbf{a} \times \mathbf{b} = -\mathbf{i} - 2\mathbf{j} + 2\mathbf{k}$ c. $\mathbf{a} \times \mathbf{b} = 3$.

   $\mathbf{a} \times \mathbf{c} = 2\mathbf{i} - 5\mathbf{y} + 2\mathbf{k}$ b. $\mathbf{a} \times \mathbf{c} = -3$

2. $x = -2$

10. (i) $(\mathbf{a} \times \mathbf{b})$ and $(\mathbf{c} \times \mathbf{d})$

    (ii) $(\mathbf{a} \times \mathbf{b}) \times (\mathbf{c} \times \mathbf{d}) = 0$

    (iii) $(\mathbf{a} \times \mathbf{b}) \cdot (\mathbf{c} \times \mathbf{d}) = 0$

11. $\mathbf{e}$ is parallel to the plane containing $\mathbf{a}$ and $\mathbf{b}$.
    $\mathbf{c}$ is perpendicular to the plane containing $\mathbf{a}$ and $\mathbf{b}$.

Exercise 14.

8. $\left|\dfrac{d\mathbf{r}}{dt}\right| = \sqrt{2}.$ $\left|\dfrac{d^2\mathbf{r}}{dt^2}\right| = 1.$

9. $\mathbf{v} = -\omega(\mathbf{a} \sin \omega t - \mathbf{b} \cos \omega t)$

   $\dot{\mathbf{v}} = -\omega^2 \mathbf{r}$

   $\mathbf{v}$ is at right-angles to $\mathbf{r}$ and $\dot{\mathbf{v}}$ is directed along the position vector towards the origin.

10. $\dot{\mathbf{r}} = -\mathbf{k}\omega \sin \omega t \mathbf{a}$

    $\ddot{\mathbf{r}} = -\mathbf{k}\omega^2 \cos \omega t \mathbf{a} = -\omega^2 \mathbf{r}$. The motion is Simple Harmonic Motion in the direction of vector $\mathbf{a}$.

11. $\mathbf{v} = \dfrac{\mathbf{r}_2 - \mathbf{r}_1}{t_2 - t_1}$

12. Speed $= \sqrt{(36a^2 + b^2)}$

13. $\mathbf{r} = \mathbf{b} + st\hat{\mathbf{a}}$

    $\mathbf{r}' = \mathbf{b}' + st\hat{\mathbf{a}}$

    $\dot{\mathbf{r}} = \dot{\mathbf{r}}' = s\hat{\mathbf{a}}$

    Hence value of $\dot{\mathbf{r}}$ is independent of choice of origin.

Exercise 14a.

1. (i) $-3\mathbf{i} - 6\mathbf{j} + 4\mathbf{k}$

   (ii) $\dfrac{x}{3} = \dfrac{y+2}{6} = \dfrac{z-6}{-4}$

2. 7**AD**

3. 20**ED**, where EB $= \frac{1}{4}$AB and BD $= \frac{2}{5}$BC.

4. 4**AC**

5. 70 m/s^2, $2\mathbf{i} + 3\mathbf{j} + 6\mathbf{k}$

6. $\frac{1}{8}\sqrt{(x^2 + y^2 + z^2)} \text{ m/s}^2$

7. (i) $\dfrac{dx}{dt}\mathbf{i} + \dfrac{dy}{dt}\mathbf{j} + \dfrac{dz}{dt}\mathbf{k}$

   (ii) $1/m \sqrt{\left[\left(\dfrac{dx}{dt}\right)^2 + \left(\dfrac{dy}{dt}\right)^2 + \left(\dfrac{dz}{dt}\right)^2\right]}$

   (iii) $1/m \sqrt{149}$

8. 0

Exercise 15.

1. 15 units.

2. 21 units.

4. 0

5. $\dfrac{\sqrt{14}}{14}(5\mathbf{i} + 10\mathbf{j} + 15\mathbf{k})$; $\dfrac{135\sqrt{14}}{14}$ units

6. Any unit vector in a plane parallel to the plane $5x + 3y + 2z = 0$.

7. $\frac{1}{3}(2\mathbf{i} + 2\mathbf{j} - 1\mathbf{k})$

Exercise 15a.

1. $\mathbf{M} = -5\mathbf{i} - 3\mathbf{j} - \mathbf{k}$. $|\mathbf{M}| = \sqrt{35}$ units

4. $\frac{5\sqrt{14}}{14}(2\mathbf{i} + \mathbf{j} - \mathbf{k})$. $|\mathbf{M}| = \frac{5\sqrt{21}}{7}$ N m

Exercise 16.

4. Radial component $\cdot54\, ak$. Transverse component $\cdot84\, ak$

5. $\dfrac{d\mathbf{r}}{dt} = k\mathbf{i} + ka \cos kt\, \mathbf{j}$

$\dfrac{d^2\mathbf{r}}{dt^2} = -k^2 a \sin kt\, \mathbf{j}$

6. (i) $\mathbf{v} = -a\omega \sin \omega t\, \mathbf{i} + a\omega \cos \omega t\, \mathbf{j}$ $|\mathbf{v}| = a\omega$

   (ii) $\hat{\mathbf{T}} = -\sin \omega t\, \mathbf{i} + \cos \omega t\, \mathbf{j}$

   (iii) $\hat{\mathbf{n}} = -\cos \omega t\, \mathbf{i} - \sin \omega t\, \mathbf{j}$

   (iv) $\mathbf{a} = \dot{\mathbf{v}} = -a\omega^2 \cos \omega t\, \mathbf{i} - a\omega^2 \sin \omega t\, \mathbf{j}$
Tangential Component 0; Normal Component $a\omega^2 = \dfrac{|\mathbf{v}|^2}{a}$.
Motion is circular, radius a.

7. (i) $\mathbf{r} = akt\, \mathbf{i} + a \sin kt\, \mathbf{j}$

   (ii) $\mathbf{v} = ak\, \mathbf{i} + ak \cos kt\, \mathbf{j}$; $\mathbf{a} = -ak^2 \sin kt\, \mathbf{j}$

   (iii) $\dfrac{-ak^2 \sin 2kt}{2\sqrt{1 + \cos^2 kt}}$

8. $-a\omega^2(2\cos\theta + 1)$, $-2a\omega^2 \sin\theta$

11. Motion is called Elliptical Harmonic Motion with axes a pair of conjugate diameters (only perpendicular if $\mathbf{a} \cdot \mathbf{b} = 0$).

12. $[(64/7M) + 5]$ units/sec

13. $\mathbf{v}_2 = 4\mathbf{i} + 7\mathbf{j} + 7\mathbf{k}$; $s_1 = \sqrt{13}$ m/s; $s_2 = \sqrt{114}$ m/s

Exercise 17.

1. $v = 4s - 3t$

3. The plane through the origin given by $3x + 4y + 2z = 0$ i.e. the plane perpendicular to the vector $(3, 4, 2)$.

4. The space (the line) generated by the vector $(2, -1, -1)$ passing through the point $(x = 2, y = -1, z = 9)$.

5. $(4, -2, 8)$.

6. $m = -1$.

12. $\begin{pmatrix} 9 & 4 & 11 & 18 \\ 8 & 2 & 6 & 12 \\ 19 & 4 & 15 & 28 \\ 8 & 3 & 7 & 11 \end{pmatrix}$

Exercise 18.

1. $\phi = 14$

$\nabla \phi = 2\mathbf{i} + 4\mathbf{j} + 6\mathbf{k}$

2. (a) $+5$

(b) $+1$

(c) $+5$

3. $\phi = 4\mathbf{i} + 4\mathbf{j} + 4\mathbf{k}$

Exercise 19.

1. (ii) $\mathbf{r} = (2\mathbf{i} + 3\mathbf{j} + 4\mathbf{k})$

(iii) $\nabla \phi = 4\mathbf{i} + 6\mathbf{j} + 8\mathbf{k}$

(iv) $2x + 3y + 4z - 29 = 0$

2. $\hat{\mathbf{n}} = \dfrac{4\mathbf{i} + 3\mathbf{j} + 8\mathbf{k}}{\sqrt{89}}$

3. $6x + 3y + 8z - 29 = 0$

4. $\theta = \cos^{-1} \dfrac{\sqrt{6}}{9}$ or arc $\cos \dfrac{\sqrt{6}}{9}$

An Introduction to Vectors

Exercise 20 (Miscellaneous).

2.
$$\begin{pmatrix} x' \\ y \\ z \\ t \end{pmatrix} = \begin{pmatrix} 1 & 0 & 0 & 0 \\ 0 & 1 & 0 & 0 \\ 0 & 0 & 1 & 0 \\ 0 & 0 & 0 & 1 \end{pmatrix} \begin{pmatrix} x \\ y \\ z \\ t \end{pmatrix}$$

3. 3540 km (approximately).

4. (i) $(\mathbf{a} \times \mathbf{b}) \times (\mathbf{c} \times \mathbf{d}) = 0$ (ii) $(\mathbf{a} \times \mathbf{b}) \cdot (\mathbf{c} \times \mathbf{d}) = 0$
    (iii) $(\mathbf{a} \times \mathbf{b}) \times (\mathbf{c} \times \mathbf{d}) \ne 0$ and $(\mathbf{a} \times \mathbf{b}) \cdot (\mathbf{c} \times \mathbf{b}) > 0$

5. Force vector $= \dfrac{F\mathbf{r}}{|\mathbf{r}|}$ $\mathbf{V}_F = \mathbf{V}_1 + \dfrac{600 F \mathbf{r}}{M|\mathbf{r}|}$

6. $(2\mathbf{i} + 3\mathbf{j} + 4\mathbf{k})$ $l^2 + m^2 + n^2 = 1$

7. $3x + 4y + 5z - 50 = 0$

8.
$$\begin{pmatrix} x' \\ y' \\ 1 \end{pmatrix} = \begin{pmatrix} 1 & 0 & a \\ 0 & 1 & b \\ 0 & 0 & 1 \end{pmatrix} \begin{pmatrix} x \\ y \\ 1 \end{pmatrix}$$

9. 0·0398 rad or $2° \ 17'$ (approximately).

10.   (i) Not possible   (ii) Linearly independent.
    (iii) $x = 1, y = 1, z = 3$ (iv) Linearly dependent.

11. $U(\frac{2}{3}, \frac{1}{3}, \frac{2}{3})$. $(\frac{2}{3}, \frac{1}{3}, \frac{2}{3})$. $6x + 6y - 9z = 0$.

12. $|\mathbf{a}|^2 - |\mathbf{b}|^2$

13. $2(\mathbf{b} \times \mathbf{a})$

14. (i) $\mathbf{a}^2 + 2\mathbf{a} \cdot \mathbf{b} + \mathbf{b}^2$ (ii) $\mathbf{0}$

15. $\mathbf{a} \times \mathbf{c} - (\mathbf{a} \times \mathbf{c}) \times \mathbf{b}$

16.   (i) perpendicular to each other.
    (ii) $|2r^2|\hat{\mathbf{n}}$ where $\hat{\mathbf{n}}$ is the unit vector perpendicular to the plane of the circle.
    (iii) $\mathbf{0}$ (iv) 0

17.   (i) $\mathbf{r} = 2\mathbf{i} + 0\mathbf{j} + 2\mathbf{k}$ (ii) $\mathbf{v} = \dfrac{5}{4\pi}\mathbf{j} - \dfrac{7}{4\pi}\mathbf{k}$
    (iii) 8 secs   (iv) $(4 + 8n)$ secs   (v) $\frac{2}{7}(1 + 4n)$ sec (n a positive integer)

18. $\mathbf{y}_1 = 707.2\mathbf{i} + 707.2\mathbf{j} + 4\mathbf{k}$

Relative velocity $\mathbf{v}_2 - \mathbf{v}_1 = -207.2\mathbf{i} - 707.2\mathbf{j} + 4\mathbf{k}$

Approximate distance $= 1474$ km apart.

19. (i) $\mathbf{v}_2 - \mathbf{v}_1$ (ii) $\mathbf{p}_1 = t\mathbf{v}_1, \mathbf{p}_2 = t\mathbf{v}_2$

(iii) distance $= t_1 \cdot |\mathbf{v}_2 - \mathbf{v}_1|$

(iv) $\mathbf{v}_3 = \dfrac{(t_1 + t_2)\mathbf{v}_2 - t_1\mathbf{v}_1}{t_2}$

20. $\mathbf{r} = \mathbf{b} + \left(\dfrac{kt^2}{2} + s't\right)\dfrac{\mathbf{a}}{|\mathbf{a}|}$ (where $\mathbf{b} = \overrightarrow{OB}$)

$\dot{\mathbf{r}} = \dfrac{kt\mathbf{a}}{|\mathbf{a}|} + \dfrac{s'\mathbf{a}}{|\mathbf{a}|}$

$\ddot{\mathbf{r}} = \dfrac{k\mathbf{a}}{|\mathbf{a}|}$

Index

Index